Mental Health in Historical Perspective

Series Editors

Aim of the Series

Covering all historical periods and geographical contexts, the series explores how mental illness has been understood, experienced, diagnosed, treated and contested. It will publish works that engage actively with contemporary debates related to mental health and, as such, will be of interest not only to historians, but also mental health professionals, patients and policy makers. With its focus on mental health, rather than just psychiatry, the series will endeavour to provide more patient-centred histories. Although this has long been an aim of health historians, it has not been realised, and this series aims to change that.

The scope of the series is kept as broad as possible to attract good quality proposals about all aspects of the history of mental health from all periods. The series emphasises interdisciplinary approaches to the field of study, and encourages short titles, longer works, collections, and titles which stretch the boundaries of academic publishing in new ways.

More information about this series at
http://www.springer.com/series/14806

Despo Kritsotaki • Vicky Long • Matthew Smith
Editors

Deinstitutionalisation and After

Post-War Psychiatry in the Western World

palgrave
macmillan

Editors
Despo Kritsotaki
Independent scholar
Athens, Greece

Matthew Smith
History of Psychiatry
University of Strathclyde
Glasgow, UK

Vicky Long
Centre for the Social History of Health and Healthcare
Glasgow Caledonian University
Glasgow, Lanarkshire, UK

Mental Health in Historical Perspective
ISBN 978-3-319-45359-0 ISBN 978-3-319-45360-6 (eBook)
DOI 10.1007/978-3-319-45360-6

Library of Congress Control Number: 2016955945

Printed on acid-free paper

This Palgrave Macmillan imprint is published by Springer Nature
The registered company is Springer International Publishing AG Switzerland
The registered company Address is: Gewerbestrasse 11, 6330 Cham, Switzerland

Contents

About the Editors and Contributors

Editors

Despo Kritsotaki is a historian of mental health and healthcare in twentieth-century Greece. Her latest projects include post-doctoral research on the Centre of Mental Health and Research (University of Crete and University of Strathclyde, 2012–14) and her work on organisations for the mental health of children and youths (THALES, University of Crete, 'Forms of Public Sociality in Twentieth Century Urban Greece', 2014–15).

Vicky Long is Senior Lecturer in History at Glasgow Caledonian University. She is the author of *The Rise and Fall of the Healthy Factory: The Politics of Industrial Health in Britain, 1914–60* (2011) and *Destigmatising Mental Illness? Professional Politics and Public Education in Britain, 1870–1970* (2014).

Matthew Smith is Professor of Health History at the University of Strathclyde. He is author of *An Alternative History of Hyperactivity: Food Additives and the Feingold Diet* (2011), *Hyperactive: The Controversial History of ADHD* (2012) and *Another Person's Poison: A History of Food Allergy* (2015).

Contributors

Leslie Baker has a PhD from the University of Saskatchewan. Her research deals with the history of mental health institutions.

Lanny Beckman has had a number of careers, including acting in director Larry Kent's pioneering Canadian film productions. In the early 1970s he founded the MPA (Mental Patients Association); from 1975 until 1990 he was editor and publisher of New Star Press; and more recently, he has participated in several projects relating to The History of Madness in Canada website (http://historyofmadness.ca/), including the MPA documentary *The Inmates Are Running the Asylum*.

Christof Beyer received his PhD in Cultural Anthropology at the University of Mainz, Germany, having studied the history of a psychiatric clinic. His main fields of research are the history of German psychiatry after the Second World War and the commemoration of the victims of the National Socialist 'Euthanasia'.

Geertje Boschma is a professor at the University of British Columbia. Her primary research focus is the history of nursing and healthcare with a specific emphasis on mental health. In addition to her involvement with the After the Asylum project (http://historyofmadness.ca/index.php?option=com_content&view=article&id=141&Itemid=142), she works on schemes related to the history of general hospital psychiatry and the transition to community mental health.

Joel T. Braslow (MD, Loma Linda University; PhD, University of California, Los Angeles) is a psychiatrist and historian at the Center for Social Medicine at UCLA. His research focuses on the ways in which sociocultural values shape and become incorporated into policy, science and clinical practice, with a focus on twentieth-century psychiatry in the United States.

Roslyn Burge is a public historian with a Master of Arts in Public History (University of Technology, Sydney), and a member of Friends of Callan Park (http://www.callanpark.com/). She has curated five exhibitions about Callan Park and initiated the Rozelle Hospital Oral History Project. Residing in Sydney, her work includes commissioned histories and oral history projects.

John Burnham is Research Professor of History at The Ohio State University, where he is also an associated scholar in the Medical Heritage Center. He has published widely on the history of psychiatry and related fields, most recently editing *After Freud Left: A Century of Psychoanalysis in America* (2012).

Megan Davies is a social historian of health with a regional specialization in British Columbia. She has researched and published on old age, social and health policy, madness, food history and home health. Since 2000 she has been involved in educational and research projects connected to the History of Madness in Canada website (http://historyofmadness.ca/) including working as film producer and co-creator of *The Inmates Are Running the Asylum* for the Mental Patients Association. Megan teaches in the Health & Society Program at York University, Toronto.

Chris Dooley is a historian who specializes in the history of mental healthcare in Western Canada and in the history of psychiatric nursing. He lives in Winnipeg where he teaches Canadian History at the University of Winnipeg, and he is a public member of the Board of the College of Registered Psychiatric Nurses of Manitoba.

Alexander Dunst is Assistant Professor of American Studies at the University of Paderborn, Germany and co-directs the research group Hybrid Narrativity: Digital and Cognitive Approaches to Graphic Literature, funded by the German Federal Ministry of Education and Research. He has co-edited *The World According to Philip K. Dick* (Palgrave Macmillan, 2015) and is the author of *Madness in Cold War America* (2016).

Erika Dyck is a Professor and Canada Research Chair in History of Medicine at the University of Saskatchewan. She is the author of *Psychedelic Psychiatry: LSD from Clinic to Campus* (2008) and *Facing Eugenics: Sterilization, Reproduction and the Politics of Choice* (2013).

Jack R. Friedman (PhD, Duke University) is a research scientist at the University of Oklahoma's Center for Applied Social Research. He has conducted anthropological research in Romania and several parts of the United States, focusing on issues related to mental health care, unemployment and responses to socio-ecological stress.

Valerie Harrington completed her Wellcome-funded PhD on the post-war history of mental health services in Manchester and Salford at the Centre for the History of Science, Technology and Medicine, University of Manchester. She currently works as a volunteer on a Macmillan Cancer support project recording life history interviews with cancer sufferers.

Nicolas Henckes is a CNRS (National Center for Scientific Research) associate researcher at the Cermes3 (Villejuif, France). He is interested in the transformation of psychiatric chronicity in the second half of the twentieth century. In his most recent project he is developing a sociological and historical analysis of concepts of psychosis risk in continental Europe.

Kathleen Kendall is Associate Professor in Sociology as Applied to Medicine at the University of Southampton, UK. Her research is broadly focused on patient and provider experiences of treatment for mental health issues.

Eugène Leblanc has been the director of the Moncton based self-help center: Groupe de support émotionnel inc, and the Publisher and Editor of Our Voice / Notre Voix since 1987. He was awarded the New Brunswick Human Rights Award in 2003.

Cheryl McGeachan is Lecturer in Human Geography in the School of Geographical & Earth Sciences at the University of Glasgow. Her research explores the historical and cultural geographies of mental (ill)health and asylum spaces. She is currently researching the historical geographies of Art Extraordinary and Scottish art therapy.

Marcia Meldrum (PhD, State University of New York) is a medical historian in the Center for Social Medicine at University of California, Los Angeles. Her research interests include the history of mental illness and mental health policy, and the treatment and management of chronic pain.

Robert Menzies is a faculty member in the Department of Sociology and Anthropology at Simon Fraser University, Vancouver, and a founding member of the History of Madness in Canada website collective (http://historyofmadness.ca/). His most recent book, co-edited with Brenda A. LeFrançois and Geoffrey Reaume, is *Mad Matters: A Critical Reader in Canadian Mad Studies* (2013).

Marina Morrow is a Professor and Chair of the School of Health Policy and Management in the Faculty of Health at York University in Toronto. In her work, Marina is interested in better understanding the social, political and institutional processes through which health and mental health policies and practices are developed and how social and health inequities are sustained or attenuated for different populations.

Howard Padwa (PhD, University of California, Los Angeles) is a historian and health services researcher with UCLA's Integrated Substance Abuse Programs. His research focuses on the formation and implementation of mental health and substance use disorder treatment policies, both in Europe and the United States.

Diane Purvey is Dean of Arts at Kwantlen Polytechnic University, British Columbia. Her research addresses notions of deviance from prescriptive norms, particularly in the context of institutions. She is the co-author of *Private Grief, Public Mourning: The Rise of the Roadside Shrine in British Columbia* (2009) and *Vancouver Noir: 1930–1960* (2011), and co-editor of *Child and Family Welfare in British Columbia: A History* (2005). A long-time resident of Vancouver's downtown eastside, Diane has engaged with residents' associations to create more inclusive communities.

Nérée St-Amand completed his PhD thesis on non-voluntary commitment and treatment of psychiatrized populations and continued to explore both the trends of psychiatric diagnosis and drugs and the multitude of unexplored alternatives to this oppressive system. For four years, he has represented families through the Mental Health Commission of Canada. He is a tenured professor at the School of Social Work, University of Ottawa, a school he created in 1990.

John Stewart is Emeritus Professor of Health History, Glasgow Caledonian University. He has published extensively on the history of health and welfare in modern Britain and is presently writing a biography of the social policy academic and Labour party advisor, Richard Titmuss.

Marie-Claude Thifault is a historian specializing in the history of asylums in Quebec and Canada. She is an Associate Professor in the School of Nursing, Faculty of Health Sciences, University of Ottawa. She is the Director of the nursing history research unit there and has been named University of Ottawa Research Chair in health and the Canadian francophonie.

Elena Trivelli received her PhD in Media and Communications from Goldsmiths, University of London in 2014. Her research examined the work of Franco Basaglia in Gorizia, with a particular focus on affect, trauma, remembering practices and forms of distributed unconscious. She currently works as a mental health advocate.

Jayne Melville Whyte is an independent author and speaker with experiential and research interest in mental health topics. She entered the mental health system in 1965 on the cusp between institutional and community services and has been active in the Canadian Mental Health Association since 1975. Jayne lives and writes in Regina, Saskatchewan.

Victor Willis is the Executive Director of the Parkdale Activity–Recreation Centre (PARC); a trustee at the Centre for Addiction and Mental Health (CAMH); a board member of Daily Bread Food Bank and chair of the Parkdale Neighbourhood Land Trust, all in Toronto. His involvement in community and mental health initiatives, ranging from public safety to housing, spans 30 years. Victor's knowledge of mental health and addiction issues are rooted in his professional, personal and family experiences.

List of Figures

CHAPTER 1

Introduction: Deinstitutionalisation and the Pathways of Post-War Psychiatry in the Western World

Despo Kritsotaki, Vicky Long, and Matthew Smith

Near the small village of Gartcosh, located in the north-eastern quadrant of the greater Glasgow conurbation, there is an imposing two-towered gothic building that used to serve as the main administration building of Gartloch Hospital. Surrounded by a fence, designed to keep people out rather than to keep them in, its windows are either broken or boarded up. Inside, what is left of the floors is strewn with detritus, ranging from broken bits of furniture and torn curtains to crumbling plaster and bent nails. It is only when one looks up to the elaborate arched and buttressed

D. Kritsotaki (✉)
Department of History, University of Crete, Heraklion, Greece
e-mail: despo.kritsotaki@gmail.com

V. Long
Centre for the Social History of Health and Healthcare, Glasgow Caledonian University, Glasgow, UK
e-mail: Victoria.Long@gcu.ac.uk

M. Smith
Department of History, University of Strathclyde, Glasgow, UK
e-mail: m.smith@strath.ac.uk

D. Kritsotaki et al. (eds.), *Deinstitutionalisation and After*,
Mental Health in Historical Perspective,
DOI 10.1007/978-3-319-45360-6_1

ceiling, painted in shades of aquamarine, scarlet and vermillion, that a hint of the former grandeur of the place becomes apparent.

Established in 1896 by the City of Glasgow and District Lunacy Board, Gartloch Hospital was one of dozens of Scottish psychiatric institutions built between the end of the eighteenth century and the middle of the twentieth century. It would exist for exactly 100 years, typically housing between 500 and 800 patients. Although it functioned primarily as a psychiatric facility for the city's poor, as with similar institutions, it also served other functions, including those of a tuberculosis sanatorium soon after it opened, and as an emergency medical services hospital during the First World War. Just prior to its closure in 1996, it was used as the set of the BBC drama *Takin' Over the Asylum*, starring Ken Stott and David Tennant.

The passage of time, alongside a renewed interest in heritage and a consequent newfound appreciation for these buildings' architectural qualities, has paved the way for redevelopment of these sites, slowly disentangling their architectural form from their erstwhile function.[1] Today, as with many former psychiatric hospitals, some of the former buildings at Gartloch have been converted into luxury apartments and the grounds are being transformed into a housing estate. On the website for Gartloch Village, ironically described as being 'far from the maddening crowd', the developers mention that a hospital was built here in 1896 (and 'immediately hailed as a Victorian architectural masterpiece'), but fail to mention its numerous psychiatric patients, focusing instead on its role treating 'returning war heroes'.[2] This approach is echoed in other asylum redevelopment projects. Friern Hospital in London, which opened to pauper patients in 1851 as Colney Hatch Asylum and at one time was England's largest asylum, closed in 1993. It reopened in 1995 as Princess Park Manor, a self-contained luxury housing development marketed as somewhere that

[1] B. Franklin (2002) 'Hospital–Heritage–Home: Reconstructing the Nineteenth Century Lunatic Asylum', *Housing, Theory and Society*, 19, 170–84. Peter Bartlett observed that the ornate features found in pauper asylum architecture reflected the survival of a charitable strand of ideology in post 1834 Poor Law thought: see P. Bartlett (1999) 'The Asylum and the Poor Law: The Productive Alliance' in J. Melling and B. Forsthye (eds) *Insanity, Institutions and Society, 1800–1914* (London and New York: Routledge), pp. 48–67. This has facilitated the repurposing of these sites from pauper institutions to luxury housing.

[2] http://gartlochvillage.co.uk/why-choose-gartloch-village/.

you 'never need to leave'. A commemorative plaque in the reception area is the only indication of the site's former purpose.[3]

While the developers of Gartloch Village may wish to obscure what the primary function of Garloch Hospital was for a century, the presence of the gloomy main administration building, its red brick tarnished by soot, tells a different story. In fact, it tells two stories. On the one hand, it reveals a Glaswegian episode in the history of psychiatric institutionalisation, a period when hundreds of thousands of people deemed to be mentally ill were housed in psychiatric asylums. But, on the other hand, the decrepit main administration building also hints at the story of what happened (and what continues to happen) next, when such institutions began to shut during the second half of the twentieth century. Unlike the history of asylums, which dominated the historiography of mental illness and psychiatry until the 1990s and continues to attract the attention of both scholars and students, the history of deinstitutionalisation is only beginning to be told. On a dreich Glasgow day, the two gothic towers of the main administration building seem to ask: How did this happen? Why did it happen? What happened to the people who used to reside behind these walls? And what happens to those today who might have found themselves behind these walls in times past?

This volume is an attempt to address some of these questions. It analyses the ideas that spurred deinstitutionalisation, charts the process of psychiatric reform in the post-war period, examines how asylum and therapeutic spaces changed for the mentally ill and assesses the ongoing legacy of deinstitutionalisation. Focusing solely on the Western world, the book explores the challenges faced by various jurisdictions in transforming psychiatric theory into practice.

Indeed, this is a theme that knits many of the chapters together. Psychiatry is a medical discipline where theories have abounded, more so than any other field of medicine. Even when these theories have begotten therapies, ranging from lobotomy and psychopharmacology to psychoanalysis and cognitive behavioural therapy, vigorous debates have followed about whether such practical applications were efficacious, ethical, cost-effective or even humane. Such was and is still the case with deinstitutionalisation. Many of the ideas that provided the framework for the closure and reformation of psychiatric hospitals were well-intended and

[3] B. Taylor (2011) 'The Demise of the Asylum in Late Twentieth-Century Britain: A Personal History', *Transactions of the Royal Historical Society*, 21, 193–215; 214.

embedded in widely accepted theory, but were problematic to apply in practice or required the sort of structural and political support that was difficult to maintain over time.

Defining Deinstitutionalisation

Deinstitutionalisation is commonly defined as the process that occurred when mental health care shifted from being based in residential asylums to becoming primarily an outpatient service during the second half of the twentieth century. Rather than living in asylums for possibly extensive periods of time, most patients were expected to live in their home communities, receiving psychiatric care and treatment in clinics and general hospitals. Others have added that the process also involved the integration of remaining psychiatric institutions and their functions with those of other health services, in order to remove the often negative consequences of institutional life on patients, such as passivity, loss of autonomy and stigma.[4] According to the World Health Organization (WHO), however, the process was meant to be more complex than mere 'dehospitalization' or 'simply reducing hospital beds'.[5] It was also intended to involve establishing high-quality care in community settings, providing comprehensive social services support to both patients and their families, and maintaining adequate acute inpatient services, as well as long-stay facilities.[6]

But deinstitutionalisation was more than merely a process of transitioning care and support from residential institutions to the community; it also constituted a significant philosophical sea change. It represented what was supposed to be a momentous transformation in the relationship between society and the mentally ill, in how mental illness was conceptualised and in how it was treated. Deinstitutionalisation cast aside the idea that the mentally ill had to be set apart from society, whether it be for their protection, society's protection (or convenience) or because secluded, secure institutions were deemed to be the most therapeutically valuable spaces for recovery. Shut away in massive institutions, such as Milledgeville, Georgia's Central State Hospital, where 12,000 patients resided as late

[4] See W. Brunton (2003) 'The Origins of Deinstitutionalisation in New Zealand', *Health and History*, 5:2, 75–103.

[5] World Health Organization (2014) *Innovation in Deinstitutionalization: A WHO Expert Survey* (Geneva: World Health Organization), p. 17, http://apps.who.int/iris/bitstream/10665/112829/1/9789241506816_eng.pdf, accessed 4 March 2016.

[6] Ibid.

as the 1960s, or solitarily in a private house, the mentally ill were cast in the role of the most undesirable 'other', a vision of humanity best kept hidden.[7] Deinstitutionalisation was meant to correct this perceived wrong. Physical reintegration was supposed to go hand in hand with a re-evaluation of the worth of people who suffer from mental illness and their full inclusion within society as citizens; the stigmatised identity of mental patient was to be discarded. In practice, this goal has only been partially achieved, demonstrating the ambiguities and inadequacies of government policy towards people suffering from mental illness, which has failed to secure their social and economic equality.[8]

For patients, the asylum, mental hospital or psychiatric institution (these terms and others will be used interchangeably in this book) would cease to be the site around which their lives literally and figuratively revolved. Instead, fortunate patients would be treated in a range of smaller and more community-oriented facilities, ranging from therapeutic communities and half-way houses to community mental health clinics and in private and public medical practices. But, as WHO and many other organisations and individuals have acknowledged, deinstitutionalisation has often amounted to dehospitalisation, with shameful repercussions. The less fortunate have not have found themselves deinstitutionalised as much as 'transinstitutionalised' into the criminal justice system.[9] A recent article in *The Atlantic* estimated that 400,000 inmates in American prisons had mental health problems, citing a National Alliance on Mental Illness estimate that between 25 per cent and 40 per cent of mentally ill Americans

[7] A. Scull (2015) *Madness in Civilization: From the Bible to Freud, from the Madhouse to Modern Medicine* (London: Thames and Hudson), p. 361; S. Wise (2012) *Inconvenient People: Lunacy, Liberty and the Mad-Doctors in Victorian England* (London: The Bodley Head); E. Said (1978) *Orientalism* (New York: Pantheon); M. Foucault (1961) *Madness and Civilization: A History of Insanity in the Age of Reason*, translated by F. Howard (New York: Vintage).

[8] P. Barham (1997) *Closing the Asylum: The Mental Patient in Modern Society* (London: Penguin); V. Long (2014) *Destigmatising Mental Illness? Professional Politics and Public Education in Britain, 1870–1970* (Manchester: Manchester University Press).

[9] Although it took post-war deinstitutionalisation for transinstitutionalisation to occur on a major scale, the British psychiatrist Lionel Penrose (1898–1972) first theorised that an inverse relationship existed between the population of prisons and asylums during the 1930s. As the population in the one increases, the population of the other decreases. L. S. Penrose (1939) 'Mental Disease and Crime: Outline of a Comparative Study of European Statistics', *British Journal of Medical Psychology*, 18, 1–15.

will find themselves behind bars at some time in their life.[10] Thousands of others have found themselves homeless, living on the streets or in temporary accommodation. In 2007, the United States Department of Housing and Urban Development (HUD) reported that 39 per cent of the nation's 600,000 homeless people reported having a mental health problem, and that between 20 per cent and 25 per cent met the criteria for serious mental illness.[11] Although the incarceration and homelessness of the mentally ill are issues of contemporary relevance, both have been recognised as consequences of deinstitutionalisation since the 1970s.[12] As had been the case in the pre-institution era, families in many countries have been left to fill in the cracks left gaping by an inadequate mental health system, providing much of the care formerly supplied by the state or private hospitals.[13]

Precursors and Drivers of Deinstitutionalisation

Deinstitutionalisation began in most countries during the years following the Second World War and reached its height between the 1960s and the 1990s, though the process has continued apace in many jurisdictions.[14]

[10] M. Ford (8 June 2015) 'America's Largest Mental Hospital is Jail', *The Atlantic*, http://www.theatlantic.com/politics/archive/2015/06/americas-largest-mental-hospital-is-a-jail/395012/, accessed 3 March 2016.

[11] U.S. Department of Housing and Urban Development Office of Policy Development and Research (2007) *The Applicability of Housing First Models to Homeless Persons with Serious Mental Illness* (Washington, DC: U.S. Department of Housing and Urban Development Office of Policy Development and Research), https://www.huduser.gov/portal/publications/hsgfirst.pdf, accessed 3 March 2016.

[12] S. Trotter and B. Kuttner (24 February 1974) 'The Mentally Ill: From Back Wards to Back Alleys', *Washington Post*; S. Stelovich (1978) 'From the Hospital to the Prison: A Step Forward in Deinstitutionalization?', *Hospital and Community Psychiatry*, 30, 618–20; G. E. Whitmer (1980) 'From Hospitals to Jails: The Fate of California's Deinstitutionalized Mentally Ill', *American Journal of Orthopsychiatry*, 50, 65–75; J. Arboleda-Florez and H. L. Holley (1988) 'Criminalization of the Mentally Ill: Part II: Initial Detention', *Canadian Journal of Psychiatry*, 33, 87–95.

[13] Scull (2015) *Madness in Civilization*, pp. 85, 121, 268–69; WHO, *Innovation in Deinstitutionalization*; W. Doll (1976) 'Family Coping with the Mentally Ill: An Unanticipated Problem of Deinstitutionalization', *Hospital and Community Psychiatry*, 27, 183–85.

[14] http://www.desmoinesregister.com/story/news/health/2015/10/08/branstads-mental-hospital-closures-debated-court/73611220/; http://www.roanoke.com/news/local/salem/salem-council-joins-opposition-to-catawba-hospital-closure/article_1e7e6a64-26e5-578a-87d4-9ddc949fab24.html; http://www.abc.net.au/pm/content/2015/s4279597.htm, accessed 15 February 2016.

It could be argued, however, that the origins of deinstitutionalisation stretch back into the era of asylum expansion in the mid-nineteenth century. Mental hospitals were criticised for being costly, inefficient and even detrimental to the well-being of patients, to which the scandalous cases of patients such as Ebenezer Haskell (1805–1892) or the intrepid investigative reporter Nellie Bly (1864–1922) attest.[15] In 1845, a group of former asylum patients in Britain established the Alleged Lunatics' Friend Society, which sought to foster public sympathies, improve asylum conditions and prevent wrongful confinement.[16] Another example of how asylums were beginning to be questioned is the French 'anti-psychiatry' campaign of the 1860s, which focused on wrongful confinement, the poor asylum and the excessive power of the psychiatric profession.[17] Critiques such as these originated from the public, the legal profession, journalists, the patients' families, and patients and former patients themselves, who in some cases played a central role in the attempts to reform asylums.

Perhaps even more importantly, deinstitutionalisation had its roots in novel community-based services that emerged in the early twentieth century, such as mental hygiene and child guidance clinics. Part of the Progressive Era social reform movements of the late nineteenth and early twentieth centuries, the American mental hygiene movement can be traced in large part to the efforts of former asylum inmate, Clifford Whittingham Beers (1876–1943), who published his asylum experiences in the autobiographical *A Mind that Found Itself* in 1908. The following year, Beers founded the National Committee for Mental Hygiene with philosopher William James (1842–1910) and psychiatrist Adolf Meyer (1866–1950). Such American endeavours were part of an international mental hygiene movement that was made up of an array of national associations, all of which, despite their distinct characteristics, criticised the conditions in psychiatric hospitals, advocated their improvement and advised treatment and aftercare outside the hospital.[18] Placing emphasis on prevention, early

[15] E. Haskell (1868) *The Trial of Ebenezer Haskell* (Philadelphia: Ebenezer Haskell); N. Bly (16 October 1887) 'Inside the Madhouse', *New York World*.

[16] N. Hervey (1986) 'Advocacy or Folly: The Alleged Lunatics' Friend Society, 1845–63', *Medical History*, 30, 245–75.

[17] I. Dowbiggin (1991) *Inheriting Madness: Professionalization and Psychiatric Knowledge in Nineteenth-Century France* (Berkeley: University of California Press), ch. 5.

[18] M. Thomson (1995) 'Mental Hygiene as an International Movement' in Paul Weindling (ed.) *International Health Organisations and Movements, 1918–1939* (Cambridge, Cambridge University Press), pp. 283–304; H. Oosterhuis (2004) 'Between Institutional

treatment and public education, mental hygiene advanced the idea that psychiatry should go beyond the hospital and engage with society, 'where things have their beginnings', as Meyer put it.[19] Although not described in terms of mental hygiene or prevention, the free clinics supported by Sigmund Freud (1856–1939) and his followers during the interwar period in Vienna, Berlin and other European cities were also inspired by the desire to provide care in the community to those who needed it most.[20] Vienna's free clinic, the Ambulatorium, reflected not only the views of Freud himself, which he expressed in 1918 at the Fifth International Congress of the International Psychoanalytical Association in Budapest, that the poor had as much right to treatment as the rich, but also fit into the socially progressive urban renewal underway in 'Red Vienna' and other parts of central Europe.[21]

The child guidance movement which emerged both in North America and Europe in close connection with mental hygiene provided an example of extramural services for children and adolescents: the child guidance clinic. In the United States (US), the child guidance movement can be largely traced to the psychiatrist and criminologist William Healy (1869–1963), who co-founded the Juvenile Psychopathic Institute in Chicago in 1909. [22] Child guidance clinics were initially connected to

Psychiatry and Mental Health Care: Social Psychiatry in the Netherlands, 1916–2000', *Medical History*, 48, 413–28; H. Oosterhuis (2005) 'Outpatient Psychiatry and Mental Health Care in the Twentieth Century. International Perspectives' in M. Gijswijt-Hofstra, H. Oosterhuis, J. Vijselaar and Hugh Freeman (eds) *Psychiatric Cultures Compared: Psychiatry and Mental Health Care in the Twentieth Century: Comparisons and Approaches* (Amsterdam: Amsterdam University Press), pp. 248–76; P. Stuart (1997) 'Community Care and the Origins of Psychiatric Social Work', *Social Work in Health Care*, 25, 25–36.

[19] Cited in K. Jones (1999) *Taming the Troublesome Child: American Families, Child Guidance, and the Limits of Psychiatric Authority* (Cambridge, MA: Harvard University Press), p. 53. For similar arguments in the case of France, see J. C. Coffin (2005) '"Misery" and "Revolution": The Organisation of French Psychiatry, 1900–1980' in Gijswijt-Hofstra et al. (eds) *Psychiatric Cultures Compared*, pp. 225–47.

[20] E. N. Danto (2005) *Freud's Free Clinics: Psychoanalysis and Social Justice, 1918–1938* (New York: Columbia University Press).

[21] Ibid, pp. 1–5.

[22] Jones, *Taming the Troublesome Child*, ch. 4; J. Stewart (2006) 'Child Guidance in Interwar Scotland: International Influences and Domestic Concerns', *Bulletin of the History of Medicine*, 80, 513–39; J. Stewart (2013) *Child Guidance in Britain: The Dangerous Age of Childhood* (London: Pickering and Chatto), p. 16; D. Thom (1992) 'Wishes, Anxieties, Play and Gestures: Child Guidance in Inter-War Britain' in R. Cooter (ed.) *In the Name of the Child: Health and Welfare, 1880–1940* (London: Routledge), pp. 189–212; and B. Evans,

courts, in order to examine juvenile delinquents, but in the 1920s and 1930s they extended beyond delinquency to treating a variety of childhood and adolescence problems rooted in personality, behaviour and relationships. In other countries, such as the Netherlands and Germany, advocates of social psychiatry, which emphasised the socioeconomic origins of mental illness, encouraged extra-mural care in the form of both preventive programmes and treatment and aftercare for outpatients in facilities such as day hospitals.[23] In Britain, outpatient consultations were introduced in the interwar period and community care for people with mental deficiencies developed as complementary to institutional care.[24]

The Second World War strengthened these discourses of mental hygiene, social psychiatry and community care, as well as precipitating broader public health and welfare measures in many countries. As the leading Anglo-Australian psychiatrist Aubrey Lewis (1900–1975) explained in 1952, four years after the foundation of the National Health Service (NHS): 'The insistent needs and the aspirations of the last war gave a vigorous impetus to social measures in every sphere. This was true of medicine in all its divisions, and notably so in psychiatry which by its nature is a field of social knowledge and practice.'[25] In the US, 12 per cent of all men who volunteered for military duty were rejected on psychiatric grounds, amounting to more than a million people. This was six times the rejection figure for the First World War.[26] Despite these high rejection rates, the American military saw over one million hospital admissions for neuropsychiatric illnesses during the course of the war. As Chief Consultant in Neuropsychiatry for the US Surgeon General, William C. Menninger (1899–1966), stated in 1948: 'Millions of people became really aware, for

R. Shahina and E. Jones (2008) 'Managing the "Unmanageable": Interwar Child Psychiatry at the Maudsley Hospital, London', *History of Psychiatry*, 19, 454–75.

[23] H.-P. Schmiedebach and S. Priebe (2004) 'Social Psychiatry in Germany in the Twentieth Century: Ideas and Models', *Medical History*, 48, 449–72; Oosterhuis, 'Between Institutional Psychiatry and Mental Health Care'; R. Dreikurs (1961) 'Early Experiments in Social Psychiatry', *International Journal of Social Psychiatry*, 7, 141–47.

[24] H. Freeman (2005) 'Psychiatry and the State in Britain' in Gijswijt-Hofstra et al. (eds.) *Psychiatric Cultures Compared*, pp. 116–40; M. Thomson (1998) *The Problem of Mental Deficiency: Eugenics, Democracy, and Social Policy in Britain c.1870–1959* (Oxford: Clarendon Press; New York: Oxford University Press), ch. 4.

[25] A. Lewis (1952) 'Forward' in M. Jones, *Social Psychiatry: A Study of Therapeutic Communities* (London: Tavistock Publications), p. vii.

[26] H. Pols and S. Oak (2007) 'War and Military Mental Health: The US Psychiatric Response in the 20th Century', *American Journal of Public Health*, 97, 2132–42.

the first time, of the effect of environmental stresses on the personality.'[27] Military psychiatry placed an emphasis on prevention (attempting to exclude from service those who were deemed prone to mental illness), adjustment (aiming to return men to battle and thus emphasising adaptation to the needs and values of patients' environment) and interpersonal treatment methods (focusing not on traditional army and hospital hierarchies but on responsibility and independence).[28] Such modifications were notable in the wartime work of some British psychiatrists, including the therapeutic community pioneer Maxwell Jones (1907–1990). As Jones explained: 'War-time needs with the huge volume of psychiatric cases and relative shortage of psychiatrists gave a tremendous stimulus to social methods of treatment in psychiatry.'[29] As the historian Catherine Fussinger describes, this included both flattening the hierarchy between patients and staff and recognising that therapy could occur not merely in a formal therapeutic setting, but 'everywhere'.[30] Wartime disorders, such as combat fatigue, supported the idea that mental illness was caused within a particular social setting and thus could be cured within this environment with early intervention and ambulatory care.

Writing about how soldiers responded psychologically to the stresses of warfare, psychiatrists Roy R. Grinker, Sr. (1900–1993) and John P. Speigel (1911–1991) described how civilians' experiences could similarly trigger mental breakdown.[31] War merely amplified phenomena which actually occurred in less fraught environments.[32] The experience of the

[27] W. C. Menninger (1948) *Psychiatry in a Troubled World: Yesterday's War and Today's Challenge* (New York: Macmillan), p. xiii.

[28] B. Shephard (1999) '"Pitiless Psychology": The Role of Prevention in British Military Psychiatry in the Second World War', *History of Psychiatry*, 10, 491–524; N. Thalassis (2007) 'Soldiers in Psychiatric Therapy: The Case of Northfield Military Hospital 1942–1946', *Social History of Medicine*, 20, 351–68; D. Healy (2002) *The Creation of Psychopharmacology* (Cambridge, MA: Harvard University Press), p. 131.

[29] M. Jones (1952) *Social Psychiatry: A Study of Therapeutic Communities* (London: Tavistock Publications), p. viii.

[30] C. Fussinger (2011) '"Therapeutic Community", Psychiatry's Reformers and Antipsychiatrists: Reconsidering Changes in the Field of Psychiatry after World War II', *History of Psychiatry*, 22, 151–52.

[31] R. R. Grinker, Sr. and J. P. Speigel, (1945) *Men Under Stress* (Philadelphia: Blakiston). For a discussion on the work of Grinker and Speigel, see M. Jackson (2013) *The Age of Stress: Science and the Search for Stability* (Oxford: Oxford University Press).

[32] Similarly, the disorder currently associated with combat stress, Post-Traumatic Stress Disorder (PTSD), which emerged specifically in the context of the Vietnam War and entered the third edition of the *Diagnostic and Statistical Manual of Mental Disorders* in 1980, has

Second World War also reinforced the psychoanalytical view that mental illness existed as part of a continuum with mental health which, in turn, had profound repercussions on how mental illness was understood, diagnosed and treated.[33] For instance, the continuum model for mental health undermined the recently ascendant biologically based explanations for mental disorder, which focused on brain pathology and reinforced the suitability of 'heroic' medical therapies, such as the use of drugs, insulin shock treatment, electro-convulsive shock treatment or lobotomy. A good example of this prior trend is the emergence of post-encephalitic disorder during the 1920s, following an epidemic of encephalitis lethargica (otherwise known as von Economo disease or sleeping sickness). In addition to suffering from movement disorders and Parkinsonism, survivors of this infectious brain disease often suffered from a wide range of mental disturbances, ranging from depression and visual disturbance to sexual precocity and violence, some severe enough to warrant lobotomies.[34] Given the obvious cause of the post-encephalitic disorder, psychiatrists were encouraged to consider similar brain lesions, dysfunctions or injuries as an explanation for other disorders.[35] Similarly, the 1906 development

more recently been associated with a range of civilian stressors, ranging from sexual abuse to witnessing the 11 September 2001 terrorist attacks on television. See V. V. Wolfe, C. Gentile and D. A. Wolfe (1989) 'The Impact of Sexual Abuse on Children: A PTSD Formulation', *Behavior Therapy*, 20, 215–28; A. Young (1995) *Harmony of Illusions: Inventing Posttraumatic Stress Disorder* (Princeton, NJ: Princeton University Press); Y. Neria, L. DiGrande and B. G. Adams (2011) 'Posttraumatic Stress Disorder Following the September 11, 2001 Terrorist Attacks: A Review of the Literature Among Highly Exposed Populations', *American Psychologist*, 66, 429–46; V. Duclos (8 October 2013) 'When Anthropology Meets Science: An Interview with Allan Young', *Somatosphere*, http://somatosphere.net/2013/10/when-anthropology-meets-science-an-interview-with-allan-young.html, accessed 15 March 2016.

[33] G. Grob (1991) *From Asylum to Community: Mental Health Policy in Modern America* (Princeton, NJ: Princeton University Press), ch. 1.

[34] F. T. Thorpe (1946) 'Prefrontal Leucotomy in Treatment for Post-Encephalitic Conduct Disorder', *British Medical Journal*, 1, pp. 312–14.

[35] E. D. Bond and K. E. Appel (1935) 'Post-Encephalitic Personality Disorders: Their Nature and Treatment', *Annals of Surgery*, 10, pp. 44–48; A. Rafalovich (2001) 'The Conceptual History of Attention-Deficit/Hyperactivity Disorder: Idiocy, Encephalitis, and the Child Deviant, 1877–1929', *Deviant Behavior*, 22, pp. 93–115; R. Mayes and A. Rafalovich (2007) 'Suffer the Restless Children: The Evolution of ADHD and Paediatric Stimulant use, 1900–1980', *History of Psychiatry*, 18, pp. 435–57. A recent article has suggested that Adolf Hitler's Parkinsonism may have been as a result of this encephalitis epidemic. K. B. Bhattacharyya (2015) 'Adolf Hitler and his Parkinsonism', *Annals of the Indian Academy of Neurology*, 18, pp. 387–90.

of the Wassermann test which identified syphilis in the blood, in tandem with the growing use of malarial therapy in the 1920s, was hailed both as a solution for general paralysis of the insane (a disease which contributed substantially to asylum death rates), but also as evidence more broadly of the biological roots of mental disorders, the scientific status of psychiatry and the utility of somatic therapies.[36] The wartime focus on the mentally pathological environment shifted the focus of many psychiatrists—albeit temporarily—away from such biological or, to use a contemporary term, organic causes and, in turn, 'heroic' medical treatments.

One of the rationales for such treatments was to ease the burden on mental hospitals, as the work of historians Jack D. Pressman and Mical Raz has demonstrated.[37] Given that one of the chief objectives of military psychiatrists was to heal psychiatric casualties sufficiently so that they could return to combat, they, too, sought short-term treatments, which, despite their limited therapeutic efficiency, advanced the idea that treatment did not demand the long-term separation of the patients from their environment.[38] Many influential military psychiatrists, such as William Menninger, urged that the presumed positive benefits of psychotherapy during wartime could and should be applied in civilian psychiatry as well, along with a host of other lessons learned during the War.[39] As historian Edgar Jones explains, this was quite different from the experience of the First World War, when, for example, the 'network of outpatient clinics, set up by the Ministry of Pensions to treat veterans with shell shock, were closed within a few years in response to financial pressures and doubts about their therapeutic value'.[40] In the US, the situation was much the same. As the anonymous reviewer of social psychiatrist Thomas A. C. Rennie's (1904–1956) *Mental Health in Modern Society* (1948) stated, 'all of the lessons we painfully "learned" during the last war [the Second

[36] Although as Gayle Davis observes, psychiatrists' subjective views of their patients and the perceived social profile of the disease mediated the impact of these new diagnostic and treatment methods: G. Davis (2008) *'The Cruel Madness of Love': Sex, Syphilis and Psychiatry in Scotland, 1880–1930* (Amsterdam and New York: Rodopi).

[37] J. D. Pressman (1998) *Last Resort: Psychosurgery and the Limits of Medicine* (Cambridge: Cambridge University Press); M. Raz (2013) *The Lobotomy Letters: The Making of American Psychosurgery* (Rochester: University of Rochester Press).

[38] E. Jones (2004) 'War and the Practice of Psychotherapy: The UK Experience 1939–1960', *Medical History*, 48, 496–97.

[39] Menninger, *Psychiatry in a Troubled World*, pp. 351–62.

[40] Jones, 'War and the Practice of Psychotherapy', 493.

World War] were well known by 1920, were published in psychiatric journals and government reports, and were tragically forgotten'.[41] After the Second World War, in contrast, psychiatrists were eager to implement such 'lessons', including the overarching notion that the mental hospital was not necessarily the best place for the mentally ill to recover.[42]

Mental hygiene, child guidance, free clinics and wartime approaches to psychiatry all help to demonstrate that alternatives to the asylum, in the form of outpatient and community care, were already present in the early twentieth century. It is also important to note that asylum walls had never been totally impermeable: there were always channels of communication between the inside and the outside world, for example in the form of patient outings, temporary leaves and boarding-out,[43] as well visits from family members, government officials, members of the press and the public.[44] Although the boundaries between the asylum and the community were not easily dissolved—as shown by the difficulties in the implementation of the formal, post-war deinstitutionalisation efforts—asylums were never completely isolated from the broader environment in which they were situated.

Origins of Deinstitutionalisation

Escalating concerns about mental health following the Second World War forced both psychiatrists and politicians to consider alternatives to hospitalisation. While mental hospitals initially remained central to the provision of mental health care and were reorganised in an attempt to improve services and to take advantage of new treatments, more emphasis was

[41] Anonymous (1948) 'Review of Thomas A. C. Rennie, Mental Health in Modern Society', *Journal of the American Medical Association*, 138, 249.

[42] Grob, *From Asylum to Community*, ch. 1.

[43] H. Sturdy and W. Parry-Jones (1999) 'Boarding-out Insane Patients: The Significance of the Scottish System 1857–1913' in P. Bartlett and D. Wright (eds) *Outside the Walls of the Asylum: On 'Care and Community' in Modern Britain and Ireland* (New Brunswick, NJ: Athlone Press), pp. 86–114; M.C. Thifault and I. Perreault (2012) 'The Social Integration of the Mentally Ill in Quebec prior to the Bédard Report of 1962', *Canadian Bulletin of Medical History*, 29, 125–50.

[44] For example, D. MacKinnon (2009) 'Amusements are provided': Asylum Entertainment and Recreation in Australia and New Zealand c.1860–c.1945' in G. Mooney and J. Reinarz (eds) *Permeable Walls: Historical Perspectives on Hospital and Asylum Visiting* (Amsterdam; New York: Rodopi), pp. 267–88.

placed on the ambulant sector and the expansion of outpatient services.[45] As a 1949 article in the American journal *Public Health Reports* indicated, there was 'widespread interest throughout the country in community mental health clinics' that would 'develop adequate mental health programs, particularly in the fields of early prevention and treatment'.[46] Such an approach was also reflected in the way in which the newly founded WHO grounded its mental health care reform policy on prevention and outpatient care.[47]

A panoply of explanations helps to explain why deinstitutionalisation happened, ranging from developments within psychiatry itself to broader political, philosophical and economic changes. The Second World War had put a spotlight on mental illness, but just what kind of psychiatry was envisioned to treat it? Post-war psychiatry was nothing if it was not diverse. While biological psychiatrists had access to new drugs, such as methylphenidate and chlorpromazine, along with other 'heroic' remedies, dynamic psychiatry was also entering its heyday, especially in the US, where being trained in psychoanalysis was a necessary condition of employment in many clinical and academic settings. At the same time, social psychiatry was highly influential, with many of the presidents of national psychiatric associations, such as the American Psychiatric Association, supporting its preventive tenets and calling for public mental health, and many politi-

[45] Stewart, '"The dangerous age of childhood"'; Jones, *Taming the Troublesome Child*, pp. 206, 210 224; M. Gijswijt-Hofstra (2005) 'Within and Outside the Walls of the Asylum. Caring for the Dutch Mentally Ill, 1884–2000' in Gijswijt-Hofstra et al. (eds) *Psychiatric Cultures Compared*, pp. 35–72; Oosterhuis, 'Outpatient Psychiatry and Mental Health Care in the Twentieth Century'; L. de Goei (1998) 'Psychiatry and Society: The Dutch Mental Hygiene Movement 1924–1960' in M. Gijswijt-Hofstra and R. Porter (eds) *Cultures of Psychiatry and Mental Health Care in Postwar Britain and the Netherlands* (Amsterdam: Rodopi), pp. 61–78; Thomson, *The Problem of Mental Deficiency*, ch. 4.

[46] E. Davens and P. Lemkau (1949) 'Birth of a Community Mental Health Clinic', *Public Health Reports*, 27, 657.

[47] S. Sturdy, R. Freeman and J. Smith-Merry (2013) 'Making Knowledge for International Policy: WHO Europe and Mental Health Policy, 1970–2008', *Social History of Medicine*, 26, 532–54. WHO's focus on prevention in the post-war period reflects its concerns today about the prevalence of mental disorders globally. Stating that depression will be the world's leading cause of disease by 2030, representing 13 per cent of global disease, WHO mentions prevention in its list of strategies to deal with the problem, but it appears to be less of a priority than in times past. WHO, 'Global Burden of Mental Disorders and the Need for a Comprehensive, Coordinated Response from Health and Social Sectors at the Country Level', Executive Board, 130th Session, Provisional Agenda Item 6.2, http://apps.who.int/gb/ebwha/pdf_files/EB130/B130_9-en.pdf, accessed 3 April 2016.

cians intrigued by its approach.[48] The launch of the *International Journal of Social Psychiatry* in 1954, which was edited by the British psychiatrist Joshua Bierer (1901–1984) and was followed by the launch of *Social Psychiatry* in 1966, also indicates its emergence as an important strand within psychiatry. Moreover, within these relatively distinct disciplines there was both diversity and overlap. Some psychoanalysts were willing to use drugs, such as Ritalin, to facilitate psychotherapy, and some biological psychiatrists, including Leon Eisenberg, who conducted some of the first clinical trials of Ritalin, were also committed to the ideas of social psychiatry.[49] Some psychiatrists who had hitherto championed physical therapies, such as Rudolf Freudenberg, one of the pioneers of insulin coma therapy, subsequently embraced social approaches and interventions.[50] Psychotherapy may have been synonymous with Freudianism to most of the lay public, but varieties ranging from those developed by Carl Jung (1875–1961) to Wilhelm Reich (1897–1957) also attracted adherents. Equally, biological psychiatry encompassed everything from shock therapies and psychosurgery to experiments in Saskatchewan to understand the therapeutic uses of LSD.[51] Even social psychiatry existed on a continuum, stretching from the relatively cautious approach of academic researchers (many of whom were social scientists and not psychiatrists) to exponents of 'radical psychiatry'.[52] These approaches to psychiatry might have been remarkably different, but they all shared one thing in common: the hopes and promises they offered all contributed to deinstitutionalisation, though to differing degrees and in different ways.

One of the defining features of dynamic psychiatry was that mental distress could affect anyone. The corollary to this was that many more people outside the asylum needed the attention of mental health care

[48] See W. C. Menninger (1949) 'Presidential Address', *American Journal of Psychiatry*, 106, 2–12; K. B. Appel (1954) 'Presidential Address: The Present Challenge of Psychiatry', *American Journal of Psychiatry*, 111, 1–12; W. E. Barton (1962) 'Presidential Address: Psychiatry in Transition', *American Journal of Psychiatry*, 119, 1–15; C. H. H. Branch (1963) 'Presidential Address: Preparedness for Progress', *American Journal of Psychiatry*, 120, 1–11.

[49] L. Eisenberg (1962) 'Preventive Psychiatry', *Annual Review of Medicine*, 13, 343–60.

[50] T. Becker and D. Bennett (2000) 'Rudolf Karl Freudenberg—from Pioneer of Insulin Treatment to Pioneering Social Psychiatrist', *History of Psychiatry*, 11, 189–211.

[51] E. Dyck (2008) *Psychedelic Psychiatry: LSD from Clinic to Campus* (Baltimore, MD: Johns Hopkins University Press).

[52] L. Richert (2013) '"Therapy Means Political Change, Not Just Peanut Butter": American Radical Psychiatry, 1968–1975', *Social History of Medicine*, 27, 104–21.

professionals. But how would dynamic psychiatrists, especially those conducting time-intensive and expensive psychoanalysis, be able to treat the countless people believed to be in need of treatment? This issue, rather than the efficacy of psychoanalysis, was the one that troubled dynamic psychiatrists during the post-war period. One solution was to open up psychoanalysis to lay analysts (those without a medical degree), something that Freud had welcomed and written about in 1926.[53] Although some psychiatrists welcomed this, stating that social workers, psychologists and even priests could perform such a function, others argued that all psychoanalysts should have a medical degree.[54] Their solution was to recruit more psychiatrists, which had long been a challenge in many jurisdictions and dominated the concerns of many psychiatric associations. Regardless of who conducted psychoanalysis, the predominance of dynamic psychiatry during the post-war period contributed to the reorientation of the psychiatric gaze from chronic and severe patients likely to be found in institutions to outpatients with acute and mild mental health problems.

Most social psychiatrists, in contrast, were not at all confident that psychoanalysis had either the efficiency or the efficacy to cope with the tide of the mentally unwell, and believed that prevention was neglected in favour of treatment.[55] Social psychiatrists believed that mental health problems stemmed from the patient's environment, often citing socioeconomic factors such as poverty, overcrowding, violence, racism and social exclusion.[56] As such, it was the community, not the hospital, that had to be the basis for both psychiatric treatment and, more importantly, prevention. Within this context more outpatient services were needed, such as the day hospitals first founded during the late 1940s by the Scot Ewen Cameron (1901–1967) in Canada and Joshua Bierer in the United Kingdom (UK), occupational mental health centres, community mental health centres,

[53] S. Freud (1990) *The Question of Lay Analysis* (New York: W. W. Norton).

[54] G. J. Sarwer-Foner (1986) 'On the Differences in the Historical Development of Psychoanalysis in the United States and Canada', *Canadian Journal of Psychiatry*, 31, 227–32.

[55] L. Eisenberg (1966) 'Discussion of Dr. Solnit's "Who Deserves Child Psychiatry: A Study in Priorities"', *Journal of the American Association of Child Psychiatry*, 5, 17–23.

[56] See R. E. L. Faris and H. W. Dunham (1939) *Mental Disorder in Urban Areas* (Chicago, IL: University of Chicago Press); A. B. Hollingshead and F. C. Redlich (1958) *Social Class and Mental Illness: A Community Study* (New York: John Wiley); L. Srole, T. S. Langer, S. T. Michael, M. K. Opler, T. A. C. Rennie, *Mental Health in the Metropolis: The Midtown Manhattan Study*, Volume 1 (New York: McGraw Hill).

therapeutic communities, specialised services for children and the elderly, follow-up services and transitional living facilities.[57] In the era of deinstitutionalisation, many of these community services were initially sited within or developed out of psychiatric hospitals, as hospitals forged new links with external organisations.[58] However, community mental health was meant to replace institutionalisation, both in terms of providing care in the community, but also by initiating the preventive strategies that would eventually see a reduction in the rates of mental illness and, therefore, the need for such facilities. Although demonstrating preventive psychiatry in practice was much more difficult than describing it in theory, it was compelling enough to convince many psychiatrists, politicians and activists that institutions could become a thing of the past.[59]

Finally, biological psychiatry's main contribution to deinstitutionalisation was the introduction of antipsychotic drugs in the 1950s. Drugs, such as chlorpromazine, sparked what has been called a 'psycho-pharmaceutical revolution', which enabled the clinical improvement of more patients and their discharge from mental hospitals, and made treating people with severe mental illnesses in the community possible.[60] Profiting from the antibiotics

[57] E. Shorter (2005) *A Historical Dictionary of Psychiatry* (Oxford: Oxford University Press), p. 59; A. Roberts and L. Kurtz (2015) 'Historical Perspectives on the Care and Treatment of the Mentally Ill', *The Journal of Sociology & Social Welfare*, 14, 75–94. Various publications on day hospitals can be found in the *International Journal of Social Psychiatry*. See for example R. A. Stewart (1956) 'The Institute of Social Psychiatry', *International Journal of Social Psychiatry*, 2, 214–19; M. Axel (1959) 'Treatment of Schizophrenia in a Day Hospital: Preliminary Observations on an Eclectic Approach', *International Journal of Social Psychiatry*, 5, 174–81 and J. Bierer (1961) 'Day Hospitals: Further Developments', *International Journal of Social Psychiatry*, 7, 148–51.

[58] D. F. Early (1963) 'The Industrial Therapy Organisation (Bristol): A Development of Work in Hospital', *The Lancet*, 281:7278, 435–36; V. Long (2013) 'Rethinking Post-war Mental Health Care: Industrial Therapy and the Chronic Mental Patient in Britain', *Social History of Medicine*, 26, 738–58. On precedents for this trend, see S. Soanes (2009) 'Reforming Asylums, Reforming Public Attitudes: J. R. Lord and Montagu Lomax's Representations of Mental Hospitals and the Community', *Family and Community History*, 12, 117–29.

[59] For a scathing indictment of social psychiatry in the US, see E. F. Torrey (2013) *American Psychosis: How the Federal Government Destroyed the Mental Illness Treatment System* (New York: Oxford University Press). Others have been more sympathetic: D. Blazer (2005) *The Age of Melancholy: 'Major Depression' and its Social Origins* (New York: Routledge); H. Pols (2007) 'August Hollingshead and Frederick Redlich: Poverty, Socioeconomic Status, and Mental Illness', *American Journal of Public Health*, 97, 1755.

[60] W. Gronfein (1985) 'Psychotropic Drugs and the Origins of Deinstitutionalization', *Social Problems*, 32, 437–54.

that proliferated after the Second World War, pharmaceutical companies sought similar magic bullets for mental illness and synthesised an array of new products, as the advertisements in psychiatry journals highlight. The new drugs however did not initially aim at overthrowing the asylum; they were firstly used in institutions to treat inpatients, regulating behaviour, facilitating hospital routine and even enabling psychoanalysis.[61] In some countries the use of antipsychotic medication coincided with the rise of psychiatric inpatients between the 1950s and the 1970s. In addition, open-door policies, community psychiatry and hospital closures had started before the advent of the first efficient drugs.[62] Therefore, although psychiatric medication contributed to the move to the community, it was not its sole or main cause and had to be complemented with social treatments in order to lead to the reintegration of patients in their social environment and to a deinstitutionalised mental health system.[63] What the new drugs, along with enthusiasm about dynamic and social psychiatry did, however, was to boost the confidence of psychiatrists that they were not only fully capable of coping with the tide of mental illness facing Western societies, but also convincing governments and policymakers to provide political and financial support for their ambitions. In the case of patients coping with severe, chronic mental illness who would have previously been institutionalised, such hopes, as we shall see, would not last. But in the case of outpatients with mild or moderate symptoms that responded well to drugs and/or psychotherapy, psychiatrists could claim success. The degree to which these more commonplace symptoms—and the new disorders that enveloped them—were the creation of pharmaceutical companies and ambitious psychiatrists, however, remains a highly contentious issue.[64]

[61] Dyck, *Psychedelic Psychiatry*, pp. 30–31. See also M. A. Ramos (2013) 'Drugs in Context: A Historical Perspective on Theories of Psychopharmaceutical Efficacy', *Journal of Nervous and Mental Disorders*, 201, 926–33.

[62] S. L. Starks and J. T. Braslow (2005) 'The Making of Contemporary American Psychiatry, Part 1: Patients, Treatments, and Therapeutic Rationales before and after World War II', *History of Psychology*, 8, 176–93; Healy, *The Creation of Psychopharmacology*, pp. 62, 129, 134; D. Pilgrim and A. Rogers (1999) *A Sociology of Mental Health and Illness* (Buckingham: Open University Press), ch. 8, table 8.1; L. Clarke (1993) 'The Opening of Doors in British Mental Hospitals in the 1950s', *History of Psychiatry*, 4, 527–51.

[63] V. Long (2012) '"Often There Is a Good Deal to be Done, But Socially Rather Than Medically": The Psychiatric Social Worker as Social Therapist, 1945–70', *Medical History*, 55, 223–39.

[64] See D. Healy (1997) *The Antidepressant Era* (Cambridge, MA: Harvard University Press); D. Herzberg (2009) *Happy Pills in America: From Miltown to Prozac* (Baltimore, MD: Johns Hopkins University Press); A. Tone (2009) *The Age of Anxiety: A History of*

Apart from the combined dynamic, social and biological strands of mental health care, deinstitutionalisation was linked to economic, cultural and political developments of the late twentieth century. The best known articulation of the socioeconomic understanding of deinstitutionalisation has been Andrew Scull's 1977 book *Decarceration*, which argued that the implementation of alternatives to institutionalisation was triggered by increasing financial pressures on the state during the 1960s and 1970s. He saw deinstitutionalisation as a 'shift in social control styles and practices' that was 'dependent upon and a reflection of more extensive and deep-seated changes in the social organization of advanced capitalist societies'.[65] Contemporary articles indicating the potential savings offered by deinstitutionalisation reinforce Scull's argument.[66] Others have pointed out, however, that the move to community care started in the 1950s and thus predated the economic strains[67] that heightened in the 1980s—the 'decade of scarcity'.[68] Indeed, Richard Warner argues that it was the demands of the post-war labour market which fuelled deinstitutionalisation, providing employment opportunities for psychiatric patients which in turn fostered the development of occupational rehabilitation services. He contends that countries which experienced higher unemployment rates after the Second World War, such as Italy, embraced psychiatric rehabilitation at a later stage, while rising unemployment levels, as in Britain from the 1970s onwards, served to curtail the development of psychiatric rehabilitation and community services.[69] In any case it cannot be denied that the conviction that community services would be inexpensive in comparison to residential ones provided an important motive to administrations for moving mental health care to the community.

The ideological and social climate of the 1960s and 1970s also provided arguments for deinstitutionalisation. Although, as already noted,

America's Turbulent Affair with Tranquilizers (New York: Basic Books); M. Smith (2012) *Hyperactive: The Controversial History of ADHD* (London: Reaktion).

[65] A. Scull (1977) *Decarceration: Community Treatment and the Deviant. A Radical View* (Englewood Cliffs, NJ: Prentice-Hall), p. 152.

[66] J. G. Murphy and W. E. Datel (1976) 'A Cost-Benefit Analysis of Community Versus Institutional Living', *Hospital and Community Psychiatry*, 27, 165–70.

[67] J. Busfield (1986) *Managing Madness: Changing Ideas and Practice* (London: Hutchinson).

[68] Roberts and Kurtz, 'Historical Perspectives on the Care and Treatment of the Mentally Ill'.

[69] R. Warner (2004) *Recovery from Schizophrenia: Psychiatry and the Political Economy* (Hove: Routledge), pp. 80–101.

the efficiency, efficacy and humanity of mental hospitals had been challenged since the nineteenth century, criticisms peaked during the decades following the Second World War. In 1959, the British psychiatrist Russell Barton (1924–2002) proposed the term 'institutional neurosis' to describe the effects of institutional life on the patients, such as apathy and lack of individuality.[70] Two years later the Canadian-American sociologist Erving Goffman (1922–1982), who had worked for the National Institute of Mental Health, characterised psychiatric hospitals as one type of 'total institutions', which cut off their inmates from wider society, restricting their self-determination, autonomy and freedom of action.[71]

Critiques such as these were accompanied by a host of others, which would attack institutionalisation from multiples perspectives. Some, like that of Burton, were from within psychiatry itself. In 1960, for instance, the Scot R. D. Laing (1927–1989) questioned why the experiences of the psychotic could not be interpreted as valid, thus questioning why such people were labelled as mentally ill.[72] Laing, who had worked at Gartnavel Royal Medical Hospital in Glasgow in the early 1950s, became convinced that the power dynamic between psychiatrists and patients in conventional mental hospitals militated against therapeutic goals, explaining to documentary maker Peter Robinson in 1971 that 'if you are interviewing a patient in a mental hospital ward and you have a key in your pocket to get out and the patient hasn't, the gulf in power, in position, is enormous'.[73] The South African psychiatrist David Cooper (1931–1986), who worked with Laing at Kingsley Hall in London and, with him and others founded the Philadelphia Association in 1965, provided a Marxist critique of the legitimacy of mental illness. He would coin the term 'anti-psychiatry', which was frequently applied to critics of psychiatry, often despite their protestations.[74] But psychiatry's critics did not have to emanate from the

[70] R. Barton (1959) *Institutional Neurosis* (Bristol: Wright).

[71] E. Goffman (1961) *Asylums: Essays on the Social Situation of Mental Patients and Other Inmates* (York: Anchor Books).

[72] R. D. Laing (1960) *The Divided Self: An Existential Study in Sanity and Madness* (Harmondsworth: Penguin); G. Miller (2004) *R. D. Laing* (Edinburgh: Edinburgh University Press); A. Beveridge (2011) *Portrait of the Psychiatrist as a Young Man: The Early Writing and World of R.D. Laing, 1927–1960* (Oxford: Oxford University Press); C. McGeachan (2014) '"The World is Full of Big Bad Wolves": Investigating the Therapeutic Spaces of R. D. Laing and Aaron Esterton', *History of Psychiatry*, 25, 283–98.

[73] 'Asylum' (1971), documentary directed by P. Robinson.

[74] D. Cooper (1967) *Psychiatry and Anti-Psychiatry* (London: Tavistock).

left. The libertarian American psychiatrist Thomas Szasz (1920–2012), for instance, was particularly critical of involuntary commitment, arguing in the falsificationist style of the philosopher Karl Popper (1902–1994)—who also had his misgivings about psychiatry—that most mental illnesses were merely metaphorical, since they could not be attributed to brain pathology.[75]

Literary and historically informed reactions to institutionalisation emerged as well. In 1962, after working at a Veterans' Hospital in Oregon, Ken Kesey (1935–2001) published *One Flew Over the Cuckoo's Nest*, which also raised questions about psychiatric institutions and 'heroic' therapies.[76] The book would later become an Academy Award-winning film in 1975. Historians, philosophers and sociologists, such as Michel Foucault (1927–1984), David Rothman and Andrew Scull, also began critiquing the asylum by examining its history, though revisionist accounts soon followed (or in the case of Grob's *The State and the Mentally Ill*, coincided).[77] Mental patients also began voicing their own opinions about institutionalisation, as the civil rights movements of the post-war period gathered momentum. By the late 1960s, the emergent Psychiatric Survivors' Movement was depicting mental patients as an oppressed minority, with mental hospitals depicted as repressive institutions that violated their human and civil rights, and psychiatry as an inhuman, authoritative or false science. Former patient Judi Chamberlain (1944–2010), who had been involuntarily confined to an asylum in New York during the mid-1960s, would articulate such concerns in *On Our Own: Patient-Controlled Alternatives to the Mental Health System*.[78] While the different social movements, organisations and individuals who contested conventional psychiatric practice frequently clashed, their interactions nevertheless served to

[75] T. Szasz (1961) *The Myth of Mental Illness* (New York: Harper and Row).

[76] K. Kesey (1962) *One Flew Over the Cuckoo's Nest* (New York: Viking).

[77] M. Foucault (1961) *Histoire de la Folie à l'Âge Classique: Folie et Déraison* (Paris: Plon); D. Rothman (1971) *The Discovery of the Asylum: Social Order and Disorder in the New Republic* (Boston, MA: Little, Brown); A. Scull (1979) *Museums of Madness: The Social Organization of Insanity in Nineteenth-Century England* (New York: St. Martin's Press); G. N. Grob (1966) *The State and the Mentally Ill: A History of Worcester State Hospital in Massachusetts, 1830–1920* (Chapel Hill: University of North Carolina Press); N. Tomes (1984) *The Art of Asylum-Keeping: Thomas Story Kirkbride and the Origins of American Psychiatry* (Philadelphia: University of Pennsylvania Press).

[78] J. Chamberlin (1978) *On Our Own: Patient-Controlled Alternatives to the Mental Health System* (New York: Haworth Press).

construct and power a field of psychiatric contention.[79] Despite their disparity, all such critics challenged madness as a medical category, proposed social models of mental distress and suggested various degrees of reform, from the transformation to the demise of psychiatric hospitals. They also, perhaps unwittingly in some cases, fuelled the desire of administrators to shut down institutions for the sake of reducing costs: if mental institutions were inhuman, oppressive and inefficient, there was no reason to spend so much money to keep them operating.[80] Pushed to the side amidst the ideology and rhetoric, however, was the issue of what would happen to the patients (both current and future) who would no longer have access to such institutions.

Different Paths and Paces

As suggested above, one could argue that deinstitutionalisation during the post-war period was continuing and building on practices already in place for at least half a century.[81] These continuities with the pre-war period notwithstanding, the scale and ideological importance of deinstitutionalisation following the Second World War situate it more suitably as a post-war phenomenon, which accelerated, generalised and intensified during the 1970s, 1980s and 1990s. Deinstitutionalisation was in no case a homogeneous process, but progressed at different paces and to different degrees in different countries and among different hospital populations. With respect to the latter, it was resisted longer and started later in institutions that cared for psychogeriatric patients, the mentally handicapped, those suffering from Down's syndrome, brain damage or other organic disabilities.[82] By and large, however, the greatest discrepancies in how deinstitutionalisation materialised occurred in different countries, not least with respect to when the process began. As with many post-war

[79] N. Crossley (2006) *Contesting Psychiatry: Social Movements in Mental Health* (Abingdon: Routledge).

[80] N. Dain (1989) 'Critics and Dissenters: Reflections on "Antipsychiatry" in the United States', *Journal of the History of the Behavioral Sciences*, 25, 3–25.

[81] Barham, *Closing the Asylum*; D. MacKinnon and C. Coleborne (2003) 'Introduction: Deinstitutionalisation in Australia and New Zealand', *Health and History*, 5:2, 1–16; Thifault and Perreault, 'The Social Integration of the Mentally Ill in Quebec'.

[82] J. Trent (1999) *Inventing the Feeble Mind: A History of Mental Retardation in the United States, 1840–1990* (Berkeley: University of California Press), pp. 241–42, 255–56, 262.

developments in mental health, the place where deinstitutionalisation got its start was the United States.

Although the American community mental health movement began in the 1940s, initiating a first wave of 'passive' deinstitutionalisation through the implementation of alternative services of mental health, it took time for the momentum behind deinstitutionalisation to build and for widespread and concerted action to occur.[83] Asylum populations continued to grow during the first decade following the end of the Second World War, rising from nearly 500,000 in 1946 to 559,000 in 1955. This was despite the emergence of a host of therapies, ranging from lobotomies and the first antipsychotic drugs to psychotherapy and therapeutic communities, believed to be capable of helping the mentally ill return to their homes. Other figures, however, reveal that a transformation was occurring. The asylum population, for example, was getting much older, a trend that had begun in the 1940s. By 1958, a third of asylum patients were over 65 years old; elderly patients also constituted an increasingly high percentage of first admissions.[84] The mental hospital was also ceasing to be where most psychiatrists worked, a profound shift from earlier eras. Whereas half of American psychiatrists worked in asylums in 1947, only 16 per cent did so in 1958, reflecting not only the decreasing reliance on institutional psychiatric care, but also the proliferation of psychotherapeutic and, to a lesser extent, research opportunities for American psychiatrists, after the foundation of the National Institute of Mental Health (NIMH) by the National Mental Health Act in 1949.[85] As the chapter on deinstitutionalisation in Canada indicates, other mental health professionals, including psychiatric nurses and social workers, were also affected by such changes.

By the 1960s, these changes were being reflected in the number of patients residing in mental hospitals. The pace varied among different states, but was generally slow between 1955 and 1965, accelerating between the late 1960s and 1970s, when deinstitutionalisation entered a more radical phase, not least because of the expansion of the welfare state which offered adequate economic support to residential alternatives.[86]

[83] Roberts and Kurtz, 'Historical Perspectives on the Care and Treatment of the Mentally Ill'.

[84] Grob, *From Asylum to Community*, p. 159.

[85] N. Hale (1995) *The Rise and Crisis of Psychoanalysis in the United States: Freud and the Americans, 1917–1985* (New York: Oxford University Press), p. 339.

[86] J. P. Morrissey (1982) 'Deinstitutionalizing the Mentally Ill: Process, Outcomes, and New Directions' in W. R. Gove (ed.) *Deviance and Mental Illness* (Beverly Hills, CA: Sage

Between 1966 and 1975, for instance, the American asylum population decreased by 250,000, reaching a total of 115,000 in 1990.[87] Making these precipitous drops more striking is that they occurred at a time when the US population was increasing rapidly, buoyed not least by the emergence of the baby boom generation, the largest cohort in American history.

Although Italy might be somewhat of an exception, in most instances, deinstitutionalisation tended to be a gradual process that required considerable political momentum. A variety of American voices, including journalist and historian Albert Deutsch (1905–1961) and journalist and mental health crusader Mike Gorman (1913–1989), had been criticising the state mental hospital system since the 1940s. Both Deutsch's *The Shame of the States* and Gorman's 'Oklahoma Attacks its Snake Pits', a pamphlet written for the National Mental Health Foundation later republished in Reader's Digest, exposed the terrible conditions, including chronic overcrowding, in state mental hospitals, generating widespread outrage.[88] One response to such concern in 1949 was the commissioning of a comprehensive study of mental health care in the then 48 states which recommended not only that overcrowding be addressed, but also that many more outpatient clinics be established.[89] Prevention of mental illness and promotion of mental health was also confirmed as a state responsibility.[90]

But, although a small number of states, including Minnesota and New York, began community mental health reform during the 1950s, and despite the formation of the National Institutes of Mental Health in 1949, presidential involvement was ultimately required to spur the federal gov-

Publications), pp. 147–76; J. P. Morrissey (1989) 'The Changing Role of the Public Mental Hospital' in D. A. Rochefort (ed.) *Handbook on Mental Health Policy in the United States* (Westport, CT: Greenwood Publishing Group), pp. 311–38; D. Mechanic and D. A. Rochefort (1990) 'Deinstitutionalization: An Appraisal of Reform', *Annual Review of Sociology*, 16, 301–27.

[87] R. Miller, A. N. Ponce, and K. S. Thompson (2011), 'Deinstitutionalization and the Community Mental Health Movement (1954–1976) in M. Rowe, M. Lawless, K. Thompson and L. Davidson (eds), *Classics of Community Psychiatry: Fifty Years of Public Mental Health Outside the Hospital* (Oxford: Oxford University Press), p. 12.

[88] A. Deutsch (1948) *The Shame of the States* (New York: Harcourt, Brace); M. Gorman (1948) 'Oklahoma Attacks its Snake Pits' (Philadelphia, PA: National Mental Health Foundation).

[89] The Council of State Governments (1950) *The Mental Health Programs of the Forty-Eight States* (Chicago, IL: The Council of State Governments).

[90] Ibid, p. 5.

ernment to get involved actively in the process of deinstitutionalisation.[91] In an address to Congress by John F. Kennedy (1917–1963) in 1963, the president lamented that 600,000 patients were 'confined and compressed' in state mental hospitals, along with 200,000 more in facilities for 'mental retardation'.[92] Kennedy, whose sister Rosemary (1918–2005) had been subjected to a botched lobotomy at the hands of the psychosurgery enthusiast Walter Freeman (1895–1972) in 1941, might have had personal reasons for his interest in mental health,[93] but the sheer financial cost of institutionalisation also played a major role in fomenting a change.[94] Kennedy calculated that $2.4 billion was being spent annually on mental hospitals, adding that this figure amounted to '$4 a day—too little to do much good for the individual, but too much if measured in terms of efficient use of our mental health dollars'.[95] Moreover, despite the fact that the American Psychiatric Association (APA) had established new standards for the quality of care and treatment in mental hospitals—essentially ensuring that they would be on par with that of general hospitals—the perception was that most institutions remained 'antiquated' and 'overcrowded'.[96] Kennedy's speech helped to pass the Community Mental Health Centers Construction Act in October 1963, which paid for the building of the community mental health centres meant to replace mental hospitals; his assassination later that year helped pave the way for an amendment which provided funding to staff these centres.

Deinstitutionalisation was often a similarly gradual—yet often convoluted—process in other countries. In England and Wales, community care was included in mental health policy since the 1930s; under the 1930

[91] American States began taking on the responsibility for caring for the indigent insane during the mid to late nineteenth century, following the campaigning of Dorothea Dix (1802–1887), which ushered in the construction of the first generation of state mental hospitals. G. N. Grob (1983) *Madness in American Society, 1875–1940* (Princeton, NJ: Princeton University Press).

[92] J. F. Kennedy (1963) 'Special Message to Congress on Mental Illness and Mental Retardation', 5 February, 1963, http://www.presidency.ucsb.edu/ws/?pid=9546.

[93] E. Shorter (2000) *The Kennedy Family and the Story of Mental Retardation* (Philadelphia, PA: Temple University Press).

[94] B. Robb (1967) *Sans Everything: A Case to Answer* (London: Nelson); A. Scull (2011) 'The Mental Health Sector and the Social Sciences in Post-World War II USA. Part 2: The Impact of Federal Research Funding and the Drugs Revolutions', *History of Psychiatry*, 22, 268–84, 271.

[95] Kennedy, 'Message to Congress'.

[96] Ibid; Grob, From Asylum to Community, p. 80.

Mental Treatment Act outpatient and voluntary treatment was allowed and encouraged.[97] The changes were codified in the 1959 Mental Health Act and mental hospitals' populations started to diminish in the mid-1950s, while community mental health centres were founded in the 1970s.[98] At the same time accusations of abuse and neglect in British psychiatric hospitals led to a string of government inquiries, which damaged the reputations of these institutions,[99] while concerns about mental hospitals was expressed during the 1960s by then Minister of Health, Enoch Powell (1912–1998), and by reformer Barbara Robb in her book *Sans Everything*, which detailed the conditions of the back wards. But, as in the US, momentum behind the transformation from institutional care to community care was slow to build: the policy of 'Care in the Community' did not get fully adopted until the 1983 Mental Health Act.[100] The government's decision to devolve responsibility for the design, funding and delivery of community-based services to local authorities, while failing to compel the establishment of such services or provide the funds necessary to do so, slowed the pace of deinstitutionalisation in England and Wales.[101] The first mental hospital closed down only in 1985, during which time mental hospitals continued to receive most of the state resources allocated to mental health. Moreover, while the number of psychiatric hospital beds decreased, the number of small hospitals and amount of short-term hospital treatment increased.[102]

[97] Freeman, 'Psychiatry and the State in Britain'.

[98] G. W. Brown, M. Bone, B. Dalison and J. K. Wing (1966) *Schizophrenia and Social Care* (London: Oxford University Press); J. K. Wing and G. W. Brown (1970) *Institutionalism and Schizophrenia: A Comparative Study of Three Mental Hospitals, 1960–1968* (Cambridge: Cambridge University Press); Pilgrim and Rogers, *A Sociology of Mental Health and Illness*, ch. 8.

[99] See, for example, National Health Service (1969) *Report of the Committee of Inquiry into Allegations of Ill-Treatment of Patients and other Irregularities at Ely Hospital, Cardiff, Cmnd. 3975* (London: HMSO); National Health Service (1972) *Report of the Committee of Inquiry into Whittingham Hospital, Cmnd. 4861* (London: HMSO).

[100] A. Scull (2004) 'The Insanity of Place', *History of Psychiatry*, 15, 417–36, 430; C. Hilton (2007) 'Changes between the 1959 and 1983 Mental Health Acts (England and Wales), with Particular Reference to Consent for Electroconvulsive Therapy', *History of Psychiatry*, 18, 217–29.

[101] K. Jones (1993) *Asylums and After: A Revised History of the Mental Health Service: From the Early 18th Century to the 1990s* (London: Athlone), pp. 181–96; T. Butler (1993) *Changing Mental Health Services: The Politics and Policy* (London: Chapman and Hall), pp. 36–51.

[102] Mechanic and Rochefort, 'Deinstitutionalization: An Appraisal of Reform'; Pilgrim and Rogers, *A Sociology of Mental Health and Illness*, ch. 8.

The passing of the 1960 Mental Health (Scotland) Act, which largely mirrored English legislation, initially suggested that Scotland would follow England's path towards psychiatric deinstitutionalisation in the 1960s.[103] Indeed Scotland was home to a number of progressive and even radical Scottish psychiatrists, ranging from George Morrison Carstairs (1916–1991) to R. D. Laing,[104] and had witnessed experiments to transform psychiatric care. Dingleton Hospital, located in Melrose in the Scottish Borders, for instance, pioneered an open-door policy in 1945 and was developed as a therapeutic community by Maxwell Jones in the early 1960s.[105] However, the Department of Heath for Scotland's insistence that psychiatric hospitals would continue to function for many years proved to be a self-fulfilling prophecy, and proposals to close Scottish psychiatric hospitals only began to be developed in the 1990s. Nevertheless, the ideologies of deinstitutionalisation effected a transformation in hospital practices, focusing attention on psychiatric rehabilitation and forging links with external organisations.[106]

The shape and extent of deinstitutionalisation differed in other countries and even within countries. In Canada, for instance, provincial jurisdiction over health meant that processes of deinstitutionalisation differed markedly in each of the ten provinces. In Saskatchewan, situated in the heart of the western prairie provinces, the pioneering socialist Co-operative Commonwealth Federation (CCF) government of Tommy Douglas (1904–1986) began reorganising mental health services soon after it was elected in 1944.[107] Such measures were part of Douglas' aim to institute a universal medical insurance system in Saskatchewan (Medicare), which began with the provision of public hospital care in 1947. By 1963, deinstitutionalisation was well under way, with the Saskatchewan Hospital in Weyburn, the largest hospital in the province (and the largest building in

[103] A. M. Keane (1987) 'Mental Health Policy in Scotland, 1908–1960' (PhD thesis: University of Edinburgh), p. 384.

[104] Carstairs' Reith Lectures in 1962 provide some insights into his views on mental health. G. Carstairs (1962) 'George Carstairs: This Island Now: The Reith Lectures', http://www.bbc.co.uk/programmes/p00h9l9f, accessed 22 March 2016.

[105] J. K. W. Morrice (1966) 'Dingleton Hospital's Therapeutic Community', *Psychiatric Services*, 17, 140–43; S. Whiteley (2004) 'The Evolution of the Therapeutic Community', *Psychiatric Quarterly*, 75, 245.

[106] V. Long (2017, forthcoming) '"Heading up a Blind Alley"? Scottish Psychiatric Hospitals in the Era of Deinstitutionalisation', *History of Psychiatry*.

[107] J. A. Mills (2007) 'Lessons from the Periphery: Psychiatry in Saskatchewan, Canada, 1944–1968', *History of Psychiatry*, 18, 179–201.

the province in the British Commonwealth at the time of its building in 1921, housing up to 2,500 patients) reducing its inpatient population by 72 per cent within five years.[108] Quebec and Alberta would soon follow in rapidly deinstitutionalising its inpatient population. Prince Edward Island, British Columbia and New Brunswick, by contrast, experienced deinstitutionalisation more gradually.[109]

Regional differences in deinstitutionalisation were also present in Australia, where states were responsible for passing their own mental health legislation and where historical, cultural and racial tensions had a profound impact on the implementation of mental health reform.[110] In contrast, New Zealand, which is more centralised, was able to carry out deinstitutionalisation further and more universally than most other Western countries, a process that began with the 1969 Mental Health Act.[111]

In Europe, experiences of deinstitutionalisation also varied. In both East and West Germany, the legacy of the Second World War weighed heavily on psychiatric services. New community-based services were introduced beginning in the 1960s, but mental hospitals remained the main locus of care until the 1980s.[112] In France, where dispensaries and 'open departments' were founded on a local and voluntary and thus limited level already in the interwar period, psychiatric reform after the war centred on the policy of sectorisation.[113] This did not aim at closing down mental hospitals, but at organising them better geographically and in relation to other

[108] Mills, 'Lessons from the Periphery', 185; C. Dooley (2011) 'The End of the Asylum (Town): Community Responses to the Depopulation and Closure of the Saskatchewan Hospital, Weyburn', *Histoire Sociale/Social History*, 44, 331–54.

[109] P. Sealy and P. C. Whitehead (2004) 'Forty Years of Deinstitutionalization of Psychiatric Services in Canada: An Empirical Assessment', *Canadian Journal of Psychiatry*, 49, 249–57.

[110] A. Rosen (2006) 'The Australian Experience of Deinstitutionalization: Interaction of Australian Culture with the Development and Reform of its Mental Health Services', *Acta Psychiatrica Scandinavica*, 113, 81–89.

[111] A. J. O'Brien and R. Kydd (2013) 'Compulsory Community Care in New Zealand Mental Health Legislation 1846–1992', *Sage Open*, http://sgo.sagepub.com/content/3/2/2158244013490175, accessed 22 March 2016.

[112] Schmiedebach and Priebe, 'Social Psychiatry in Germany in the Twentieth Century'.

[113] N. Henckes, 'French Deinstitutionalisation or the Irony of Success: Psychiatrists, the State and the Transformation of the French Psychiatric System, 1945–2010', ch. 6 in this volume.

services.[114] As Nicolas Henckes describes in his chapter, however, deinstitutionalisation occurred nevertheless despite the ambitions of psychiatrists and the state to improve psychiatric hospitals. In the Netherlands, since the 1960s, new outpatient services were available but they expanded to meet the needs of patients with serious disorders only in the 1980s and 1990s, when large psychiatric hospitals were downsized.[115] In Finland, the process similarly occurred later, but rapidly, as the population in Finnish mental hospitals reduced by one third during the 1990s.[116] As Elena Trivelli describes in her chapter, Italy went further than all other countries, with Law 180 in 1978 beginning the 20-year process during which all mental hospitals were shut. In Greece there was an early, albeit weak, interest in community psychiatry since the late 1940s, but apart from a few isolated initiatives of community and outpatient treatment, psychiatric care was mainly provided in mental hospitals and private clinics until the 1980s. At that time deinstitutionalisation took off, but has been impeded by various factors, such as a lack of trained personnel, extensive bureaucracy, mismanagement of funds and more recently, with the economic crisis, the decrease in public funding.[117]

Outcomes

The process of deinstitutionalisation has been appraised in different ways since the 1980s. Most commentators tend to agree that, although in many countries it has resulted in a decentralised, 'pluralistic mental health sector funded by a diversity of public and private programs', it has not meant the end of the mental hospital.[118] On the contrary, in some countries, such as the US, the foundation of community services did not reduce rates of admission to public hospitals.[119] The implementation of deinstitutionalisation programmes

[114] Coffin, '"Misery" and "Revolution": The Organisation of French Psychiatry, 1900–1980'.

[115] Oosterhuis, 'Between Institutional Psychiatry and Mental Health Care'.

[116] R. K. Salokangas and S. Saarinen (1998) 'Deinstitutionalization and Schizophrenia in Finland: I. Discharged Patients and their Care', *Schizophrenia Bulletin*, 24, 457–67.

[117] A. Blue (1999) *The Making of Greek Psychiatry* (Athens: Exantas), pp. 119–25, in Greek and M. Mitrosyli (2015) *Leros Mental Hospital and Reform. Public Policies, Institution, Community* (Athens: Papazisi), in Greek.

[118] Mechanic and Rochefort, 'Deinstitutionalization: An Appraisal of Reform', quote on p. 322.

[119] Healy, *The Creation of Psychopharmacology*, p. 150; G. N. Grob (1996) 'The Severely and Chronically Mentally Ill in America: A Historical Perspective' in S. M. Soreff (ed.)

demonstrated the ongoing requirement for institutional care, as residential units and hospitals had specific functions which could not easily be replaced or removed.[120] Indeed what has often happened was not deinstitutionalisation—the removal of patients from institutions—but trans-institutionalisation—the transfer of patients to alternative institutions, such as nursing homes, correctional institutions, private mental hospitals, general hospitals, board-and-care homes and other residential facilities.[121]

Few service users mourn the demise of the old asylums.[122] Nevertheless, for those who were truly deinstitutionalised, leaving institutions to live in the community, support has most often been inadequate.[123] One reason for this has been the limited planning and resources for services and professional help in the community, but families and communities have also been opposed to deinstitutionalisation, arguing that at least some of the mentally ill were unable to live outside the institutions and/or posed a threat to the community.[124] Exacerbated by cuts to the social welfare programmes in many countries that began with the rise of neo-liberal governments during the 1980s in North America, the UK and elsewhere, community care was increasingly perceived not as professional care delivered in the community but as informal, voluntary and unpaid care by the community, namely by relatives, friends and neighbours, care that was not always possible or optimal. As a result, the mentally ill were at greater risk of lacking the specialised treatment they needed, of losing genuine opportunities for kinship, friendship, and neighbourhood networks and of facing

Handbook for the Treatment of the Seriously Mentally Ill (Seattle, WA: Hogrefe & Huber), pp. 27–44.

[120] J. E. Gudeman and F. S. Miles (1984) 'Beyond Deinstitutionalization: A New Class of Facilities for the Mentally Ill', *New England Journal of Medicine*, 311, 832–36.

[121] G. Grob (2005) 'The Transformation of Mental Health Policy in Twentieth-Century America' in Gijswijt-Hofstra, Oosterheis, Vijselaar and Freeman (eds) *Psychiatric Cultures Compared*, pp. 141–61; Roberts and Kurtz, 'Historical Perspectives on the Care and Treatment of the Mentally Ill'; Mechanic and Rochefort, 'Deinstitutionalization: An Appraisal of Reform'; E. Shorter (1997) *A History of Psychiatry: From the Era of the Asylum to the Age of Prozac* (New York: John Wiley & Sons), p. 281.

[122] See, for example, P. Campbell (1992) 'A Survivor's View of Community Psychiatry', *Journal of Mental Health*, 1, 117–22; Barham, *Closing the Asylum*, pp. 39–65. Diana Gittens' work indicates that the closure of some asylums was greeted with trepidation. D. Gittens (1998) *Madness in its Place: Narratives of Severalls Hospital, 1913–1997* (London: Routledge).

[123] Roberts and Kurtz, 'Historical Perspectives on the Care and Treatment of the Mentally Ill'.

[124] Fox, 'Debating Deinstitutionalisation'.

the prospect of homelessness or incarceration.[125] Homelessness, in particular, could cause vicious cycles of episodic hospitalisations and unsupported discharges, creating significant problems for the administration of mental health services and adding to the already pronounced stigma faced by the mentally ill, as it strengthened the public perception of the mentally disordered as vagrant and possibly dangerous individuals.[126] These limitations have led some to claim that 'deinstitutionalization remains an unfulfilled promise'[127] or that 'the true "shame of the states"' was not the asylum, as Albert Deutsch had contended in 1948 for the United States, but deinstitutionalisation.[128] Indeed, the psychiatrist and historian Trevor Turner has argued that 'a gradual return to more institutional provision seems likely', citing the backlash against community care fuelled by stories of perceptions of the risk posed by some individuals who suffer from mental illness.[129] However, this contention is hard to square with the continuing decline in the number of psychiatric inpatient beds, and even if deinstitutionalisation—at least in the way it was implemented—has failed in many respects, its predecessor, the asylum/residential hospital, has also been irreversibly discredited.

Contents

Deinstitutionalisation and After: Post-War Psychiatry in the Western World[130] is divided into four parts, each examining deinstitutionalisation from a different perspective. Part I, 'Understanding Deinstitutionalisation: Culture, Ideology and Historiography' begins with John Burnham's argument that for deinstitutionalisation to occur, society had to undergo a 'tectonic shift' from a culture made up of producers, to one consisting of consumers. Prior to deinstitutionalisation, asylums had been the spaces

[125] C. Fox (2003) 'Debating Deinstitutionalisation: The Fire at Kew Cottages in 1996 and the Idea of Community', *Health and History*, 5, 37–59; P. Horden and R. Smith (1998) 'Introduction' in P. Horden and R. Smith (eds) *The Locus of Care: Families, Communities, Institutions, and the Provision of Welfare Since Antiquity* (London: Routledge), pp. 1–18.

[126] H. R. Lamb and L. L. Bachrach (2001) 'Some Perspectives on Deinstitutionalization', *Psychiatric Services*, 52, 1039–45.

[127] Mechanic and Rochefort, 'Deinstitutionalization: An Appraisal of Reform', 324.

[128] Shorter, *A History of Psychiatry*, p. 280.

[129] T. Turner (2004) 'The History of Deinstitutionalization and Reinstitutionalization', *Psychiatry*, 3:9, 1–4, 4.

[130] Our title nods to the work of noted British historian of psychiatry Kathleen Jones (1922–2010), particularly Jones, *Asylums and After*.

that housed society's unproductive members. The development of consumer culture in the post-war Western world meant that patients no longer needed to be productive; they did, however, have to be consumers. Burnham's chapter, which draws on literature about consumer culture, illustrates how deinstitutionalisation not only reflected a monumental change in psychiatric care, but also deep-seated transitions in terms of how an individual's contribution to society was measured.

Whereas Burnham's chapter deals with mainstream consumer culture and its relationship to deinstitutionalisation, Alexander Dunst analyses how radical ideas about psychiatry also informed the transformation of post-war psychiatry. Dunst's chapter examines American community magazines which were published by ex-patients and sympathetic mental health workers during the 1970s and early 1980s. Dunst argues that the magazines served a number of purposes to those who contributed to them and the thousands of people who read them. On the one hand, they helped to articulate the desires of radical therapists, many of whom saw patient liberation only in social transformation or revolution. But, on the other hand, the magazines were also therapeutic spaces in their own right, fuelling a more introverted self-help philosophy that distanced itself from politics.

Concluding the first part, the chapter by Megan Davies and Erika Dyck et al. emphasises how deinstitutionalisation in Canada similarly had unintended consequences. Espousing a bottom-up approach that relies on the testimony and experiences of both patients and mental health practitioners, the authors portray deinstitutionalisation in Canada as a complex, contradictory process that varied across the ten provinces. Particularly notable is the authors' willingness to reflect upon how their project changed as they constantly sought to understand how deinstitutionalisation was 'experienced directly'. As a result, their chapter is as much about the methodology of mental health history and how to capture and value the voices of the voiceless as it is about an important episode in the history of Canadian mental health.

Part II, 'Designing and Implementing Psychiatric Reform: Experiments, Opportunities and Oppositions' turns to four case studies of how deinstitutionalisation materialised in Europe. Christof Beyer begins by exploring the psychiatric reforms of Karl Peter Kisker at the Hannover Medical School during the 1960s and 1970s. Inspired by the social psychiatric reforms in the US and the UK, Kisker and his colleagues attempted to establish a community mental health approach, but their ambitious plans were undermined by the difficulties inherent in treating the most needy patients and the continued reliance on hospitals as the centre of 'community care'.

In France, as Nicolas Henckes explains, post-war mental health policy was dictated, as elsewhere, by concerns about escalating rates of mental illness, but unlike many other countries, the mental hospital was viewed as part of the solution rather than part of the problem. The policy of sectorisation, which was similar in nature to the catchment area policy in other jurisdictions, was intended to help improve and integrate outpatient services for the mentally ill, while retaining the hospital as the hub for such services. But, while hospitals continued to be constructed, the integration of services was left incomplete, resulting in 'an incredibly complex and fragmented landscape of institutions and services', rendering many of the most vulnerable 'invisible'.

Addressing a more focused case study, Valerie Harrington explores how community mental health services developed in Salford, Lancashire, arguing that it represented a rare example of how a local authority embodied the principles of the 1959 Mental Health Act (England and Wales). Seeking to elucidate why Salford was unique in this respect, Harrington examines the 'minutiae of change', ranging from the experiences, motivations and interactions of protagonists, such as Lance Burn, Salford's Medical Officer of Health, and the relevant social, political and geographical context. In so doing, she makes a compelling case for ensuring that local cases, such as Salford, inform grander narratives about deinstitutionalisation.

The final chapter of this section turns to Greece, where Despo Kritsotaki demonstrates how the Centre for Mental Health and Research helped to make the case for deinstitutionalisation long before the process actually began in the 1980s. In addition to emphasising the impact of the Centre in gradually laying the groundwork for deinstitutionalisation by working with communities, Kritsotaki highlights how patients' families influenced the implementation of community mental health care, in some cases, obstructing it. Deinstitutionalisation eventually did occur in Greece, but it was a negotiated process that was dependent on changing social attitudes and the willingness of patients and families to accept it as a new approach to mental health care.

In Part III, 'New Conceptualisations of Therapy and Space', two novel therapeutic approaches that emerged in post-war Britain are considered. John Stewart assesses how concerns about the 'maladjustment' of British children stimulated preventive approaches, most notably, child guidance. While children from dysfunctional backgrounds might previously have been removed from their home and institutionalised, Stewart describes how this policy was abandoned during the post-war period. Instead, psychiatrists, social workers and psychologists became convinced that they could re-educate parents and, thus, prevent or reverse the maladjustment of their children. Some aspects of child guidance were increasingly called into question: for example viewing the family, rather than socioeconomic factors, as the locus of pathology. Nonetheless, Stewart concludes that child guidance contributed to the demise of children's institutions.

Turning to a very different approach to mental illness, Cheryl McGeachan focuses on Kingsley Hall, the experimental community founded by R. D. Laing and Aaron Esterson during the 1960s. Employing the methodologies of geography, McGeachan analyses the sort of therapeutic space Laing and Esterson envisioned and what it actually became for the patients who lived there. McGeachan argues that such 'alternative conceptualisations of therapeutic "spaces" (both real and material, imagined and envisioned) matter in the exploration of the deinstitutionalisation process'. Such experiments were fraught with both practical and ethical challenges, may have been dismissed at the time and may have led to tragedies in some instances, but nevertheless represented a bold attempt to change thinking about mental illness and the therapeutic process, something that is worth remembering.

We conclude in Part IV, 'After Deinstitutionalisation: Experiences, Challenges and Legacies' by presenting the aftermath of deinstitutionalisation in Italy, California and Australia. Elena Trivelli begins this section with her analysis of Italian deinstitutionalisation, specifically the work and legacy of Franco Basaglia. Law 180, which was passed in 1978 and is commonly known as 'the Basaglia Law', led to the replacement of Italian mental hospitals with community-based services. Focusing on Trieste, one of the cities in which Basaglia was based, Trivelli uses ethnography to argue that the way in which Basaglia's legacy has been invoked, mythologised and 'crystallised' runs counter to the way it should be interpreted, ending the revolution he sought to instigate rather prematurely.

In our penultimate chapter, Howard Padwa, Marcia Meldrum, Jack Friedman and Joel Braslow consider how deinstitutionalisation has been experienced by three different mental health stakeholders in California: mental health administrators and providers, family members of the mentally ill and the mentally ill themselves. For administrators and providers, the adoption of neo-liberal principles meant that community mental health care became as much about inculcating independence and autonomy as it was about fostering support and enabling therapeutic relationships. In response, family members reacted by forming advocacy groups that provided mutual support and lobbied to fight the stigma of mental illness and pushed for better public mental health services. Although these two stakeholders found promise in California's recovery model of mental health services, the authors' interviews with patients revealed that simple lack of support and care made it very difficult for them to gain the empowerment, independence and community connections required for recovery. Despite many attempts to reform and deinstitutionalise mental health services in California since the early 1970s, many of the problems remain, leaving patients struggling to receive the help they need.

With our final chapter, from Roslyn Burge, we almost come full circle. In it, she describes the process by which the Australian government repeatedly tried to shut down Callan Park, a mental hospital in New South Wales, eventually succeeding in 2008 in the hope of selling it off for profit. This was in spite of the protestations of local community residents who valued the hospital and advocated the expansion of mental health services. Rather than desiring deinstitutionalisation, the Friends of Callan Park desired improvements in institutional care, including making the hospital more integrated with the community in which it was enveloped.

By exploring how psychiatric institutions might have been reimagined, rather than replaced wholesale, Burge's chapter leaves us wondering about the lessons left by deinstitutionalisation. While most psychiatrists and patients would not welcome a return to a reliance on institutions, there may be a place for such facilities, however reimagined and reconfigured. The term 'asylum' may seem archaic and pejorative, but, during a time when discussion of 'asylum-seekers' dominates the news agenda in many Western countries, perhaps we should reconsider what the word can also convey—a place of refuge. Community care might be preferable in the long term, but it is possible that people facing intractable psychiatric problems still need what asylums offer at difficult times in their life. Reminiscing about her time as a patient in Friern Hospital in the late 1980s, the historian Barbara Taylor describes how 'I could no longer do ordinary life, and giving up the struggle was an incalculable relief… Living in the bin was tough, but it gave me some shelter from my darkest self and, very importantly, the friendship of other patients.'[131] While acknowledging the deficiencies of psychiatric hospital care, Taylor lambasts the 'indi-

[131] Taylor, 'The Demise of the Asylum', pp. 212–13.

viduated and disconnected' nature of community care. 'The much-touted independence of the community-based user', she observes, 'often equals a life of lonely isolation.'[132]

Equally, although deinstitutionalisation has been mercilessly attacked by its many critics, including the psychiatrist cum historian E. Fuller Torrey, it is likely that such assessments focus far too much on how deinstitutionalisation was carried out in practice, rather than the ideals upon which it was based.[133] The high rates of serious mental illness amongst prisoners and the homeless is a terrible indictment of how the promoters of deinstitutionalisation failed to provide the support required by countless psychiatric patients as the asylums emptied out and in the time since then. But such failures were not due to innate faults in the idea of deinstitutionalisation, but the inability and unwillingness of policymakers, politicians and the leaders of the mental health establishment to carry the process out as fully as was needed. Similarly, the notion that mental illness may be prevented seems just as elusive as it was in 1945, but if as much resource and energy was dedicated to identifying and eliminating the causes of mental illness as is invested in the development of new psychiatric drugs, perhaps prophylactic strategies would not be quite the mirage they are often described to be.

In compiling *Deinstitutionalisation and After: Post-War Psychiatry in the Western World* we aim to inspire more historians to ask searching questions about how ideas about mental health and psychiatry have evolved during the turbulent decades since the Second World War.[134] We hope that, like many of the authors in this volume, they not only describe what has happened in the field of mental health, but also take the steps to inform debates about how policy and practice might improve in future. We admit that it has been impossible to cover all the themes, debates, places and people who shaped and were shaped by deinstitutionalisation, but encourage others to take up this task in the belief that this history will add immeasurably to how we understand and experience mental health and illness today and how we might improve the lives and prospects of the mentally ill in future.

[132] Ibid, p. 215.

[133] Torrey, *American Psychosis.*

[134] This reflects the recent call by Greg Eghihan to 'deinstitutionalize' the history of psychiatry, that is, by exploring the history of psychiatry and mental health outside the asylum. G. Eghihan (2011) 'Deinstitutionalizing the History of Contemporary Psychiatry', *History of Psychiatry*, 22, 201–14.

PART I

Understanding Deinstitutionalisation: Culture, Ideology and Historiography

CHAPTER 2

Deinstitutionalisation and the Great Sociocultural Shift to Consumer Culture

John Burnham

Introduction

Historians have found many factors that brought about the destruction of mental institutions in the decades after the Second World War. Some studies focus on local circumstances. Some use particular hospitals as illustrative cases of broader trends. Others trace regional or national policy-making, sometimes to the level of individual influential historical figures.[1] Yet a basic question remains: why did so many patients leave so many hospitals in so many different settings within the time frame of a single generation?

This chapter is designed to call attention to a broad sociohistorical change within which deinstitutionalisation of mental patients took place, more general perhaps than the decarceration movement, for

[1] One very recent list of possible causative factors is found in G. Paulson (2012) *Closing the Asylums: Causes and Consequences of the Deinstitutionalization Movement* (Jefferson, NC: McFarland & Company), especially ch. 2.

J. Burnham (✉)
Ohio State University, Columbus, OH, USA
e-mail: burnham.2@osu.edu

D. Kritsotaki et al. (eds.), *Deinstitutionalisation and After*,
Mental Health in Historical Perspective,
DOI 10.1007/978-3-319-45360-6_2

example, named by Andrew Scull.[2] An essential background factor, shared worldwide, was the shift from a producer culture to a consumer culture. That shift was specific for shaping the ways in which European peoples changed how they handled their mentally ill fellow citizens in the second half of the twentieth century.

Despite the fact that deinstitutionalisation took place under different political and social circumstances in different parts of the world, the tectonic sociocultural change to a consumer culture came into place at some time before, or simultaneously with, deinstitutionalisation. As the details worked out, deinstitutionalisation depended substantially upon changing assumptions about personal identity and the relationship of an individual to his or her immediate social setting. My main conclusion, then, is that in the pervasive new consumer culture, deinstitutionalisation could seem to be natural and make sense to people in the cultures in which it happened.

The best way to understand what took place is first to look at the institutions for mental patients in the late nineteenth and early twentieth centuries, and then to look at what happened to mental patients in the second half of the twentieth century—essentially 'before' contrasted with 'after'. I therefore shall begin by recalling to readers how institutions of the late nineteenth and early twentieth centuries were overwhelmingly based on the producer culture of those times so that the contrast with later decades becomes clear, that is, the picture 'before' and the picture 'after'. Indeed, I expect that many readers are already acquainted with my 'before' stage as well as at least parts of my 'after' stage. The 'after' stage, however, may be more difficult for us to see because we are still so surrounded and driven by it—not least in the ways in which it has affected the most severely ill mental patients.[3]

[2] A. T. Scull (1977) *Decarceration: Community Treatment and the Deviant: A Radical View* (Englewood Cliffs, NJ: Prentice-Hall). Mental patients/mentally ill people generally consisted of those who were developmentally disabled, brain damaged, or mad (later 'psychotic') as measured by local standards.

[3] Evidence of unawareness is offered in T. Kasser and A. D. Kanner (2004) (eds) *Psychology and Consumer Culture: The Struggle for a Good Life in a Materialistic World* (Washington, DC: American Psychological Association).

The Development of Work-Oriented Mental Institutions

At the end of the eighteenth and beginning of the nineteenth centuries, the number of mental hospitals increased noticeably, and they did so in cultures that were becoming increasingly bourgeois.[4] Those who devised and operated the hospitals were, at their best, focused on welfare and humanitarian goals. The motives of those setting up asylums were always mixed, as indicated by the governor whose 1766 message brought about the first exclusively mental hospital in the British North American colonies. After speaking of humanitarian considerations, he described 'a poor unhappy set of People who are deprived of their Senses and wander about the Country, terrifying the rest of their Fellow Creatures ... A legal Confinement and proper Provision, ought to be appointed for these miserable Objects, who cannot help themselves.' And then he added, 'Every civilized Country has an Hospital for these People.'[5]

As the institutions became medicalised with physicians in charge, the goal of curing the patients became more prominent. Many times, those physicians mentioned work or occupation as a part of the therapy.[6] 'Exercise, occupation, and amusement' were simply part of managing an asylum. 'All three are in turn requisite and indispensable, though not all are equally required by the same individual', declared G. Fielding Blandford in his lectures at St George's Hospital in London as late as 1870.[7]

In different locations, managers of asylums included work by both men and women as part of 'moral treatment', perhaps most famously at Wakefield in the UK. Historian Leonard D. Smith recounts how one asylum manager after another found that work could be not only therapeutic but also useful in reducing costs by making the institution at least partially self-supporting. Earlier, William Tuke at the pioneering York Retreat

[4] In many cases the asylum was a humanitarian alternative to the poorhouse of that day, where the point was that inmates were expected to work.

[5] N. Dain (1971) *Disordered Minds: The First Century of Eastern State Hospital in Williamsburg, Virginia, 1766–1866* (Charlottesville: University Press of Virginia), p. 7.

[6] J. Laws (2011) 'Crackpots and Basket-Cases: A History of Therapeutic Work and Occupation', *History of the Human Sciences*, 24, 65–81, offers a historiographical background and overview of work as therapy.

[7] G. F. Blandford (1976 [1871]) *Insanity and Its Treatment: Lectures on the Treatment, Medical and Legal, of Insane Patients* (New York: Arno Press), p. 380.

had held work the most 'efficacious' element in treatment.[8] Wilhelm Griesinger, the eminent psychiatric writer in Germany in the 1860s and 1870s, still similarly declared, 'Of moral remedies proper, employment of the patient seems to us the most important.'[9]

From the middle to the late nineteenth century, two major developments changed the nature of the institutions. First, it became obvious that neither hospitalisation nor moral treatment was curing patients. Second, increasing numbers of patients came flooding into the hospitals. Managers therefore shifted much of their attention away from curing patients and instead concentrated on caring for patients. From the mid-nineteenth century through to the First World War, in the Anglophone countries and in continental Europe, first one government and then another launched a great building programme to accommodate the large numbers of patients who came crowding into mental hospitals. In those large institutions, moral treatment and work therapy tended to become obscured simply by the task of managing such large numbers. The great achievement of the asylums was to house the mentally ill so that they would simultaneously be protected and would not interfere with the normal functioning of society. The major problems the managers faced were overcrowding, physical disease and the least amount of mechanical restraint they could employ for control.[10] Many medical experts began to think that most patients would in fact not be cured.[11] Gradually, in this new warehousing asylum era at the end of the nineteenth century, medical superintendents once again began to discover empirically the value of inmate labour. In Scotland, for example, as one long-time witness reported, 'It became evident that asylums required land on which to employ suitable patients … [also] the district asylums soon developed the agricultural ambition, and attained success in

[8] L. D. Smith (1999) *'Cure, Comfort and Safe Custody': Public Lunatic Asylums in Early Nineteenth-Century England* (London: Leicester University Press), ch. 7. A. Scull (1993) *The Most Solitary of Afflictions: Madness and Society in Britain 1700–1900* (New Haven, CT: Yale University Press), quotation from p. 102.

[9] W. Griesinger (1882) *Mental Pathology and Therapeutics*, trans. C. L. Robertson and J. Rutherford (New York: William Wood and Company), p. 346.

[10] See especially P. McCandless (1979) '"Build! Build!" The Controversy over the Care of the Chronically Insane in England, 1855–1879', *Bulletin of the History of Medicine*, 53, 553–74; Scull, *The Most Solitary of Afflictions*, 303–15; M. A. Arieno (1989) *Victorian Lunatics: A Social Epidemiology of Mental Illness in Mid-Nineteenth-Century England* (Selinsgrove, PA: Susquehanna University Press).

[11] See, for example, E. Shorter (1993) *A History of Psychiatry, From the Era of the Asylum to the Age of Prozac* (New York: John Wiley & Sons), ch. 3, on Germany.

the undertaking.' Patients who worked anywhere on the grounds or in everyday functioning improved their own morale and the morale of the institution. But that work still supported the institution financially, often very significantly.[12]

In the face of having to keep overwhelming numbers of the mentally ill in a culture in which a work ethic was strongly promoted by leadership elements, psychiatrists began to develop new lines of reasoning that helped shape work programmes, beyond the considerations of morale, order and distraction. If cure was not a major, achievable goal, there was an alternative to the recovery model: the rehabilitation model. That is, handicapped people perhaps could not be cured, but they could be trained to carry out some productive activity, that is, work. The rehabilitation model became particularly effective around the time of the First World War, when physically injured soldiers were trained to re-enter the workforce in spite of their handicaps.[13] So in the asylums, the inmates, once they were productive, could be prepared to join the regular workforce or at least could be productive enough to help sustain the institution. Either way, the inmate/patient would have at least some of the self-respect an individual would have in a producer-dominated society. And, once again, there was the same discovery that had been made many decades earlier: work helped patient morale.

By the end of the nineteenth century, then, explicit discussions of the importance of work for institutionalised patients were increasing in much of the psychiatric literature. Particularly in the first three decades of the twentieth century, the managers of one hospital after another extended patients' working into the regular routines of the establishment. In part, of course, this was more possible because the steep class divisions of an earlier time were being overwhelmed by the influx of masses of patients from

[12] H. M. Hurd (1916–1917) (ed.) *The Institutional Care of the Insane in the United States and Canada* (4 vols., Baltimore, MD: Johns Hopkins University Press), especially 1: 242–249: 'During the period from 1870–1890 there was a large development of indoor industries in connection with the various state institutions', in addition to out-of-door and farm work (1: 249). A. R. Urquhart (1910) 'Lunacy Administration in Scotland, With Special Reference to the Royal Asylums', *Journal of Mental Science*, 56, 626. And see the summary in K. Jones (1991) 'The Culture of the Mental Hospital' in H. Freeman and G. E. Berrios (eds) *150 Years of British Psychiatry, 1841–1991* (London: Gaskell), pp. 17–28.

[13] B. Linker (2011) *War's Waste: Rehabilitation in World War I America* (Chicago, IL: University of Chicago Press), explicitly ties the rehabilitation programme to the work ethic of that period. See also D. Cohen (2001) *The War Come Home: Disabled Veterans in Britain and Germany, 1914–1939* (Berkeley: University of California Press).

all walks of life. In 1828, George Man Burrows had complained that 'The greatest difficulty is to find occupation or amusement for the higher classes of lunatics. They sooner get tired of the same pursuit. Reading, billiards, chess, cards, and other games, must be diversified for in-door—walking, bowls, gardening, and athletic exercises, for out-door—amusement.'[14] At the turn of the twentieth century, not only did work become labelled as 'therapy', but even many of the 'higher class' of patient could be appealed to through the bourgeois dedication to the work ethic that suffused the culture.

The Work Ethic

The work ethic had come into Europe from a variety of sources and always contained some internal contradictions. Yet it was held in various forms across Europe as part of middle-class ideology and varied in influence as the bourgeois elements were able to exert cultural influence. It was (as famously suggested by social thinker Max Weber) most strident among Protestant populations. In the mid-nineteenth century, moralists in England such as Thomas Carlyle who praised the value of being constantly productive had particular influence not just in Britain but in the industrialising north of the United States, where the work ethic was most vociferously advocated and where middle-class groups had become culturally dominant. The idea was that working and producing was virtuous and morally imperative, and that hard work would pay off in gaining 'independence' and 'success' for oneself.[15]

In each culture, a unique set of institutions and beliefs embodied the high value put on work, but some form of the work ethic existed throughout European areas. It was, for example, a belief shared by most voluntary emigrants to North America from all over Europe, an easily observed phenomenon showing how universal the assumption was that productivity was not only a moral imperative but brought a promise of a better living. Moreover, scholars have shown that what eventually happened was that the process of work, not the product of the work, was what became simultaneously morally

[14] G. M. Burrows (1976 [1828]) *Commentaries on the Causes, Forms, Symptoms, and Treatment, Moral and Medical, of Insanity* (New York: Arno Press), p. 707.

[15] D. T. Rodgers (1979 [1978]) *The Work Ethic in Industrial America, 1850–1920* (Chicago, IL: University of Chicago Press), especially ch. 1; H. Applebaum (1992) *The Concept of Work: Ancient, Medieval, and Modern* (Albany: State University of New York Press), especially ch. 18.

and socially valuable. One did not need to attain one's goals; one could be happy just striving for them.[16]

The extent to which the producer culture affected psychiatrists can be shown in the general adoption of the ability to work as a test for mental illness. Indeed, it should not be surprising, since failure to work was a deviance from the surrounding bourgeois culture. The standard may be best known in the often-cited dictum of Sigmund Freud that mental health meant the ability to work and to love.[17] Specifically, in the late nineteenth and early twentieth centuries, a person's ability to be productive was a major test for mental deviance and for a person's ability to exist outside of an institution. In endless case after case in psychiatric reports, inability to work meant the patient was ill, and when the patient had gained the ability to perform productively so as to contribute to society, that was a sign that the patient had recovered and could rejoin the community.

Moreover, many institutions became organised so that the hospital community would, ideally, be self-sufficient in operations because of the labour of many patients. This was possible with a rural population in which men were used to farming and women did such domestic chores as cleaning, cooking and sewing. In the minds of those in charge of mental hospitals, all such activities were considered therapeutic as well as productive. For the individual patient, there was an important sense of being useful and in fulfilling social goals. At the same time, hospital chores helped train patients for work if and when they were released.

Martin Barr of the Pennsylvania Training School in 1899 spelled out some of the assumptions as they applied to the developmentally disabled:

> all true happiness must have its root in self-respect .... To avoid pauperizing and to stimulate self-respect[,] we must offer some avenue for gaining the respect and appreciation of others. This is best done by uplifting and maintaining the dignity of labor, and this for all grades [of feeble-minded]. Even the lowest is made to feel: 'Who sweeps a room as to thy law, makes that and the action fine'.[18]

[16] Rodgers, *The Work Ethic*, especially pp. 271–73. Indeed, some thinkers commented on the frantic activism of the culture; Applebaum, *The Concept of Work*, p. 460.

[17] See D. Riesman (1950) 'The Themes of Work and Play in the Structure of Freud's Thought', *Psychiatry*, 13, 1–16.

[18] M. W. Barr (1899) 'The How, the Why, and the Wherefore of the Training of Feeble-Minded Children', *Journal of Psycho-Asthenics*, 4, 207–208.

Eventually, in many places, formal programmes of occupational therapy were developed, still looking to the ultimate release of the patient as well as the immediate support of the institution in terms of both productivity and morale. It was hard to draw lines. For example, in the era before antipsychotic drugs, depressed patients seemed to benefit from repetitious tasks. Thus polishing the floor of the ward, over and over, as shown in Fig. 2.1, was gratifying activity for some patients.

The work ethic even operated to the extent that mental hospitals began to change the rationale when administrators tried to provide organised recreation as an adjunct to work, for this all took place at a time when uplifting recreation was supposed to build character and develop discipline. And, it should be added, the hope was that with that discipline, a patient would better conform to the factory system and be productive if the individual were able to be released from the hospital.

The work ethic lost dominance across both classes and countries in the decades after 1950 (as the factory system also diminished relative to service industries in advanced economies). Meanwhile, beginning in the 1950s, deinstitutionalisation proceeded to develop as a movement. Perhaps the most striking example of the simultaneous and interacting end of institutional work and rise of deinstitutionalisation came from hospitals influenced by

Fig. 2.1 Patients polishing the hospital floor, 1930s (author's collection)

the famous German model established by the work of psychiatrist Hermann Simon of Gütersloh. Just as the First World War was coming on, he developed his 'aktiv' work therapy, inspired in part by the rehabilitation model. This system influenced hospitals not only in Germany but throughout at least northern Europe. In the late 1920s, there were on average each year more than a dozen official visitors from outside Germany formally recorded at the Gütersloh asylum, in addition to many score from sister German institutions. The Simon patient work system lasted from the 1920s until the 1960s. What caused it to fade away was the deinstitutionalisation movement in Germany and also in other countries, particularly the Netherlands, which rendered the system obsolete and irrelevant.[19] That is, as the culture shifted, there was no longer conceptual support in the 1960s and after for having institutionalised patients work and thus support the existence of the institution.

After: Consumer Culture

By the end of the twentieth century, it was possible to maintain that the only function an individual in developed countries needed to perform to be a full citizen and a complete person was to consume. It had therefore become conceivably possible for people whose behaviour, from at least one point of view, was extremely deviant to live outside of an institution in a way that would fulfil the requirements to be a person and a free citizen. I turn therefore to the coming of consumer culture and how deinstitutionalisation could not have proceeded without that fundamental shift—stipulating again that, parallel to deinstitutionalisation, the shift from a producer culture to a consumer culture varied globally in extent, timing and manifestation. It also came on gradually. Already in the late 1920s, Simon in Germany was noticing that some of the younger schizophrenics

[19] F. C. Stamm (1975) 'The Netherlands' in J. G. Howells (ed.) *World History of Psychiatry* (New York: Brunner/Mazell), p. 163. See especially B. Walter (1996) *Psychiatrie und Gesellschaft in der Moderne: Geisteskrankenfürsorge in der Provinz Westfalen zwischen Kaiserreich und NS-Regime* (Paderborn: Ferdinand Schöningh), especially pp. 253–67; F.-W. Kerstling (2003) (ed.) *Psychiatriereform als Gesellschaftreform: Die Hypothek des Nationalsocialismus und der Ausbruch der Sechziger Jahre* (Paderborn: Ferdinand Schöningh). Simon is treated in detail from a particular point of view in H. L. Siemen (1987) *Menschen Blieben auf der Strecke ...: Psychiatrie zwischen Reform und Nationalsocialismus* (Gütersloh: Verlag Jakob van Hoddis). H. Simon (1927) 'Aktivere Krankenbehandlung in der Irrenanstalt', *Allgemeine Zeitschrift für Psychiatrie*, 87, 97–145, etc.

and mentally deficient objected to working. They came to the institution to rest and recover ('zur Erholung') and thought therefore that they should not have to work.[20]

Able historians have described the coming of modern consumer culture, from Thorstein Veblen's 1899 description of 'conspicuous consumption' that characterised 'the leisure class', to the late twentieth century, when the primary psychological identity of the typical citizen tended to be that of a consumer, not a worker, and when both the classes and the masses pursued consumption for its own sake. In many details, with local contingencies, this cultural change was uneven, and many strong traces of the producer culture of course lasted into the consumer era. Cultural change is never clean and clear so as to satisfy philosophers. There were and are many exceptions. Nevertheless, for historians, general trends like the coming of consumer culture should be discernible, even with deinstitutionalised patients—or perhaps especially with deinstitutionalised patients.

One impression of what had happened was recorded as early as 1982 in New York City by Gerald Weissman. Alluding to a number of famous clothiers there, he connected consumer culture to the homeless people who filled formerly unremarkable public spaces. He described a

> … subclass of 'deinstitutionalized' mental patients who have congregated in our cities in a kind of behavioral mockery of the consumer society … the sad battalions of mad ladies course our streets, loaded with possessions tucked in tattered paper bags. And the bags still carry their message of fashionable competition: BLOOMINGDALE'S! BONWIT TELLERS! BERGDORF GOODMAN! The bags, in anarchic disorder, are in turn stashed in that other symbol of our abundant life: a shopping cart from the supermarket. There is probably no other country in which the madmen and madwomen so neatly exhibit the claims of local enterprise. I cannot recall similar public displays of private goods among the cat ladies of Rome, the mendicants of Madrid, the down-and-out of London and Paris. Madness is surely as prevalent in those cities as in New York, but mockery of THE SHOPPER in the person of the bag lady strikes me as pure Americana …. It has always seemed to me to constitute a fantastic notion that the social landscape of our large cities bears any direct relationship to that kind of stable, nurturing community which would support the fragile psyche of the mentally ill. Cast into an environment limited by the welfare hotel or park bench, lacking adequate outpatient services,

[20] Simon, 'Aktivere Krankenbehandlung', p. 107.

> prey to climatic extremes and urban criminals, the deinstitutionalized patients wind up, as often as not, conscripts in an army of the homeless.[21]

It was of course, regardless of Weissman's impressions, a scene repeated, using local specifics, in urban areas throughout the developed world.

Scholars have described how for decades a plenitude of goods, distributed through new establishments like department stores, advertised and publicised relentlessly in the mass media (particularly at first in newspapers and magazines and then movies and radio), gradually came to dominate at least middle-class life in the Euro-American world. The new culture overwhelmed first one social group and then another. The most spectacular shift occurred late in the twentieth century amongst populations in former Iron Curtain countries who turned to preoccupation with consumer goods—most notably the ingress, initially, of conspicuous status items like personal radios, watches and types of clothing. And then of course, automobiles.

For some time in the second half of the twentieth century, the consumer culture was so pervasive in most areas that for a long time people did not see it as such. Only in recent decades have social analysts recognised it.[22] They have generally shown that a person entrapped in a dominating consumer culture after about 1950 might have two fates: to be manipu-

[21] G. Weissman (1982) 'Foucault and the Bag Lady', *Hospital Practice*, 17 (8), 29, 33. All of the stores mentioned were at the time upscale clothing retailers with major outlets in New York City.

[22] The earlier treatments tend to be the most cogent. The standard works that established this point of view for the United States include D. M. Potter (1954) *People of Plenty: Economic Abundance and the American Character* (Chicago, IL: University of Chicago Press); W. I. Susman (1984) *Culture as History: The Transformation of American Society in the Twentieth Century* (New York: Pantheon Books); R. W. Fox and T. J. J. Lears (eds) (1983) *The Culture of Consumption: Critical Essays in American History, 1880–1980*). The best summary is J.-C. Agnew (1990) 'Coming Up for Air: Consumer Culture in Historical Perspective', *Intellectual History Newsletter*, 12, 3–21. Comparable works are being produced for all Westernised societies, and there is a scholarly analytic *Journal of Consumer Culture*. J. Baudrillard's 1970 book, *The Consumer Society: Myths and Structures* [*La Société de Consummation*], trans. C. T. (London: Sage Publications), was translated into English only in 1998. Once the subject of consumer culture became recognised, scholars with political interests flooded into the field and wrote about goods and sociopolitical movements and in the process missed the fact that it was a whole culture as such. Many recent writers who did write about consumer culture as such were fundamentally interested in marketing and even the business practicalities of marketing. See, for example, D. J. Goodman and M. Cohen (2004) *Consumer Culture: A Reference Handbook* (Santa Barbara, CA: ABC-Clio); C. Lury

lated, exploited and victimised on the one hand, or, to be empowered and developed by the ability to make choices and to make choices that would enhance one's individuality and sense of personhood on the other hand.[23] Deinstitutionalisation of mental patients in practice accorded with both aspects of consumer culture.

Deinstitutionalisation as Adaptation to Consumer Culture

Like consumer culture, the deinstitutionalisation movement also came at different times in different societies. The conspicuous effects of deinstitutionalisation did not appear in Japan until the 1990s. But in each case, a developing consumer culture made the change feasible. Over many decades in the twentieth century, then, there emerged a new social place for psychotic, developmentally disabled, and brain-injured patients. Obviously that place was outside of mental institutions where they could patronise vendors in a variety of marketplaces.[24] A new rhetoric, taken from the surrounding consumer culture, gradually overshadowed the old role and reflected the new one that emerged for the mentally disabled person as a consumer. It was as consumers that the disadvantaged were supposed to achieve their humanity in that new era.[25]

One dramatic demonstration of the transformation is the way in which occupational therapists adjusted their activities and goals in the second half of the twentieth century. It was particularly after the First World War that occupational therapy as a special field became established. Then during the

(1996) *Consumer Culture* (New Brunswick, NJ: Rutgers University Press); R. W. Belk and J. F. Sherry, Jr. (2007) (eds) *Consumer Culture Theory* (Oxford: Elsevier).

[23] See, for example, the summary by S. Thornham (2000) *Feminist Theory and Cultural Studies: Stories of Unsettled Relations* (London: Arnold), ch. 6.

[24] General background is in such books as K. Jones (1993) *Asylums and After: A Revised History of the Mental Health Services: From the Early 18th Century to the 1990s* (London: Athlone Press), and the works of G. N. Grob, such as (1991) *From Asylum to Community: Mental Health Policy in Modern America* (Princeton, NJ: Princeton University Press).

[25] In the 1970s, it was possible to argue that mental patients in their role as consumers should replace 'citizens' in community mental health governance: W. E. Holton, P. K. New and R. M. Hessler (1973) 'Citizen Participation and Conflict', *Administration in Mental Health*, 2, 96–103. In the 1990s, the drive was for empowerment of the consumers of mental health services, or at least feelings of empowerment: E. S. Rogers et al. (1997) 'A Consumer-Constructed Scale to Measure Empowerment among Users of Mental Health Services', *Psychiatric Services*, 48, 1042–7.

mid-century decades, deinstitutionalisation forced the field to change.[26] In general, occupational therapists moved to the so-called community, along with the patients. By the 1980s, occupational therapists had long been focusing on independent living, on money management and on life skills. Those skills consisted largely of how to make consumer choices and spend money. Consuming was what deinstitutionalisers meant by full participation in the community. Full participation did not include work. Rather, participation and personal fulfilment came, ideally, from shopping. Patients were supposed to adapt to the community, but now it was a community devoted to consumer culture and largely informed by commercial advertising. It is ironic that people bearing the title of occupational therapist were taking patients out to learn how to shop and buy, not to work, not even in the service industries that became predominant.

It is easy to see how consumer culture standards that affected psychiatric patients drew upon many complementary currents already present by the middle decades of the twentieth century. One current was the idea of social adjustment, a concept common by the 1930s in psychiatry, psychotherapy and psychology. That is, a patient was encouraged to adapt to his or her social circumstances, and a person's inability to adapt was counted as a sign of having mental problems; as the author of one transitional book put it, the person was 'socially incompetent' in meeting 'the demands of the outside world'.[27] Indeed, one major thinker of that era, the Swiss-American psychiatrist, Adolf Meyer, defined mental illness as failure to adjust, which could show itself as social maladaptation as well as inability to work.[28] It was but a small step to generalise that social adaptation was desirable behaviour (people even today make fun of 'plays well with others'). By the mid-century decades, it was beginning to be a consumer society to which one obviously had to adapt to show one's mental health, not the discipline of a producer society. The change came inexorably, mixing the old with the new. For example, one therapy team as late as 1973 still

[26] D. Bennett (1996) 'Work and Occupation for the Mentally Ill' in H. Freeman and G. E. Berrios (eds) *150 Years of British Psychiatry, Volume II: The Aftermath* (London: Athlone Press), pp. 193–208.

[27] S. P. Davies (1930) *Social Control of the Mentally Deficient* (New York: Thomas Y. Crowell), p. 376.

[28] M. Raz (2010) 'Psychosurgery, Industry and Personal Responsibility, 1940–1965', *Social History of Medicine*, 23, 117.

thought that some sort of employment was a good sign, but they did not rate it as highly as a 'full social life'.[29]

This consumer thinking was essential in the arguments of an important element in the new psychiatric and social strategy, the normalisation movement. Patients were now supposed to be placed in normal, not institutional, environments. They were also supposed to follow normal patterns of living which, it turned out, consisted of shopping and buying, and leisure and social activities. Productivity was not an essential goal of normalisation. Rather, mentally ill people were learning to live, socialise and consume.

In lieu of a family home or an institution, the group home was supposed to constitute independent living for the mentally impaired, or it was, in lawyers' formulations, the 'least restrictive' environment for those people. But it was clear that the group home was dominated by buying and consuming. No vocational skill or even ability in everyday living was necessary for deinstitutionalised people. As one deinstitutionaliser explained, 'Purchase of gourmet frozen foods, laundry, and other services ensures that they can live comfortably in private dwellings.'[30] Even socialising typically involved shopping and buying, for example acquiring a suitable appearance with clothes and beauty aids, or 'going out', typically dating and courting in places of commercial, that is, purchased, amusement.

'Independent living', noted Gerben DeJong in 1978, 'has become a code word for deinstitutionalisation and for the establishment of community based alternatives.' The goal became, as independent living advocate Gareth Williams put it, 'participation in the world, whether or not that participation included a vocational component'.[31] Indeed, in the independent

[29] A. Pam et al. (1973) 'Community Adjustment of Self-Discharged Patients', *Psychiatric Quarterly*, 47, 179.

[30] B. Lozano (1993) 'Independent Living: Relation Among Training, Skills, and Success', *American Journal on Mental Retardation*, 98, 249–62; the quotation is from p. 261. The paradoxical observation that successfully living independently did not even necessarily depend upon learning living skills led to the articulation of this point of view.

[31] DeJong, 'The Movement for Independent Living'; quotations are from p. 2. Williams, 'The Movement for Independent Living', pp. 1003–1010; the quotation is from p. 1003. It is ironic that, as P. L. Tyor and L. V. Bell (1984) *Caring for the Retarded in America* (Westport, CT: Greenwood Press), pp. 123–27, point out, the initial impetus for returning patients to the community (once they were no longer considered dangerous) was based on vocational 'rehabilitation'. J. W. Trent, Jr. (1994) *Inventing the Feeble Mind: A History of Mental Retardation in the United States* (Berkeley: University of California Press), chs 4 and 6, gives more detail and a different twist to the story.

living literature, it was clear that strong supervision was necessary for patients placed in group homes or apartments. The independence was in fact only rhetorical, of course. Empowerment was a matter of labelling and perception.

After: Consumerism and Choices

By the end of the twentieth century, then, productivity as a driving force in life was fading out of the world of the mentally ill, institutionalised or deinstitutionalised, as it was in most surrounding populations. Of course some people worked conscientiously to fulfil the role of consumer, but now their duty consisted of thinking about and choosing goods and services in their minds, whether or not they actually bought anything. That is, the measure of normality was the extent to which they responded dutifully to advertising and sales efforts in their environments. As summarised in an article by Canadian professionals in the *Scandinavian Journal of Occupational Therapy* in 1999, as in the rest of the population, the preoccupation of both therapists and mental patients had become the 'subjective quality of life', as opposed to the 'usefulness' of an earlier day.[32]

The extreme of expectation was that mentally ill people were supposed to choose and purchase their own medical and psychological treatments, just as any sick person was supposed to shop for health care among multiple 'providers'. Indeed, Dylan Ronald Tomlinson in the UK argued that if mental patients could choose and purchase medical and other services, they would *not* lose status as they would if the services were rendered free on the basis of handicap. Uncontrolled choice was very important in a 'marketplace' where all consumers (or clients, not 'patients') were equal. 'Many potential customers might feel more comfortable', wrote Tomlinson, 'where agencies were not aiming at supervision of their minds.' This curious linkage in rights, between 'freedom' from both institutional discipline on the one hand, and welfare entitlement (along with the right, not just to services, but to choose services) on the other hand,

[32] G. Aubin, R. Hachey and C. Mercier (1999) 'Meaning of Daily Activities and Subjective Quality of Life in People with Severe Mental Illness', *Scandinavian Journal of Occupational Therapy*, 6, 53–62. For patients still in institutions, the economic model of normality showed up even in a commonly used therapeutic token economy, in which money equivalents were used to reward desirable behaviour. The tokens, the centre of the normalisation training, were to be used to purchase goods and services.

epitomised the consistent connection between deinstitutionalisation and consumer culture.[33]

That is, in consumer culture, the ultimate of self-fulfilment and personhood was making commercial decisions. Choice was the ultimate human attribute in both the free market consumer society of Thatcher, Reagan and various social theorists, for example, and the normalisers, who believed that those with mental difficulties should be encouraged to take chances, suffer consequences and not be sheltered as if they were different kinds of people.[34]

It was no accident that many of the most enthusiastic advocates of hospital closures were entrepreneurs who offered commercial choices and stood to make money off the mentally ill, such as, notoriously, operators of private nursing homes or group homes.[35] In the United States, at least, free-market extremism as a part of consumer culture brought many terrible things to deinstitutionalised mental patients—isolation, homelessness, exploitation and imprisonment.[36]

[33] See, for example, A. Crosson (1977) (ed.) *Advocacy and the Developmentally Disabled* (Eugene: College of Education, University of Oregon); D. R. Tomlinson (1991) *Utopia, Community Care and the Retreat from the Asylums* (Milton Keynes: Open University Press). I am not dealing here with a whole category of ethical arguments that emerged, including distributive justice and equality of persons; see such works as E. M. Bower (1957) 'Cultural Values and the Retarded Child', *Mental Hygiene*, 41, 201–206; L. R. Tancredi (1977) *Ethical Policy in Mental Health Care: The Goals of Psychiatric Intervention* (New York: William Heinemann Medical Books); R. B. Potter (1977) 'Labeling the Mentally Retarded: The Just Allocation of Therapy' in J. Reiser, A. J. Dyck and W. Curran (eds) *Ethics in Medicine: Historical Perspectives and Contemporary Concerns* (Cambridge, MA: MIT Press), pp. 626–31. N. Tomes (2006) 'Patients or Health-Care Consumers? Why the History of Contested Terms Matters' in R. A. Stevens, C. E. Rosenberg and L. R. Burns (eds) *History and Health Policy in the United States: Putting the Past Back In* (New Brunswick, NJ: Rutgers University Press), pp. 83–110, discusses how the patient was turned into a 'consumer' and the doctor into a commercial 'provider' in healthcare in general, not just in psychiatry.

[34] See, for example, H. M. Leichter (1991) *Free to Be Foolish: Politics and Health Promotion in the United States and Great Britain* (Princeton, NJ: Princeton University Press).

[35] B. C. Vladeck (1980) *Unloving Care: The Nursing Home Tragedy* (New York: Basic Books). Paulson, *Closing the Asylums*, describes the greed of developers who wanted hospital real estate, but instances are mentioned repeatedly in many other accounts, not only in North America, but in Europe, where government bureaucrats often displayed acquisitive desire for hospital land and buildings.

[36] R. J. Isaac and V. C. Armat (1990) *Madness in the Streets: How Psychiatry and the Law Abandoned the Mentally Ill* (Arlington, VA: Treatment Advocacy Center). Paulson, *Closing the Asylums*, offers another informed view.

Ironically, most of those who argued for non-institutional care in the community were unaware of the powerfully and socially conservative effect of a consumer culture, spelled out in the classic essay on consumerism by Jean-Christophe Agnew. Shopping involves making choices in one's imagination. Because one does not need money, anyone can make those imaginary choices.[37] Consumer culture thus brought to society a false equality, an equality of dreams, without disturbing the existing social and economic hierarchy in that society. Deinstitutionalisation in a consumer culture was still one more guilt-free way for citizens and their legislators to avoid providing fundamental protection to the most helpless people in a society.[38]

The nature of the disability of severely ill mental patients has always been precisely in the area of making choices: their perceptions were askew or their ability to process perceptions into actions were undeveloped or damaged. Yet consumer culture depended upon a person's making choices, particularly in a world of media, overt advertising and shopping. People who made errors in choice had only themselves to blame.[39] Obviously mental patients, responding to cultural cues, were to be held responsible for their decisions and actions. No institution with medical services existed to shelter them from the consequences of their disabilities. Instead, they had the advantages of a free market in which, like anyone else, they could choose between advertised brands and goods and activities depicted on television.[40] It was in this context that we see the programmes of the occupational therapists and social workers to try to train patients to shop and still feel good about themselves because they were participating in consumer culture.

[37] J.-C. Agnew (1983) 'The Consuming Vision of Henry James' in Fox and Lears (eds) *Consumer Culture*, pp. 65–100. Baudrillard, *The Consumer Society*, also makes the point. Cultural impact is spelled out in detail in R. Marchand (1985) *Advertising the American Dream: Making Way for Modernity, 1920–1940* (Berkeley: University of California Press).

[38] As S. Miles (2012) 'The Neoliberal City and the Pro-Active Complicity of the Citizen Consumer', *Journal of Consumer Culture*, 12, 216–30, points out, consumption-dominated modern cities did indeed reflect the free-market neo-liberalism of the turn of the twenty-first century, including, we can add, the deinstitutionalisation of the newly mentally ill who greatly swelled the ranks of the homeless.

[39] For a detailed analysis of choice in a consumer culture, see for example Y. Gabriel and T. Lang (1995) *The Unmanageable Consumer: Contemporary Consumption and its Fragmentation* (London: Sage), ch. 2.

[40] P. Lerman (1982) *Deinstitutionalization and the Welfare State* (New Brunswick, NJ: Rutgers University Press) made another argument, that only in countries with welfare systems where one was entitled to economic support on the basis of citizenship could the consumer culture function, so there was no question about having to be productive and work. Otherwise consumer choice worked only for those with money—hopefully lots of money, which would apply to large parts of populations in developed countries.

Conclusion

In 1981, when medication compliance had become probably the major issue in trying to deal with deeply afflicted patients in the community, psychiatrists in various locales could report that coffee groups—groups centred around acts of consumption—could be used as instruments to try to keep very sick patients on their antipsychotic pills, or at least to choose a medication and choose to take it. The point, of course, was to keep deeply disturbed people out of hospitals or, by that time, gaols. Otherwise, what one London psychiatrist in 1988 described as the whole programme of 'pushing psychiatric services out of the ageing Victorian asylums' could be discredited by excessive amounts of socially disruptive behaviour.[41]

The fact that traditional mental hospitals had been based on the work ethic of another age was a major factor in undermining their social function in the post-Second World War decades. My conclusion is therefore easily stated: the great social shift to a consumer culture in the second half of the twentieth century was a necessary—but as other chapters in this volume show—not a sufficient cause of deinstitutionalisation.[42]

[41] S. W. Olarte and R. Masnik (1981) 'Enhancing Medication Compliance in Coffee Groups', *Hospital and Community Psychiatry*, 32, 417–19; T. Turner (1988) 'Community Care', *British Journal of Psychiatry*, 152, 1; never mind that the appeal to possess new, rather than 'aging', facilities was of course a commonplace of consumer culture.

[42] This paper is based on a chapter in my unfinished book on deinstitutionalisation, and I have borrowed much documentation and some wording from that chapter. Many colleagues, in a variety of settings, have offered comments and suggestions on that chapter and on the argument in this paper, and I am grateful to each and every one of those kind people. A draft of this particular paper was critiqued by the writing workshop of the Medical Heritage Center of Ohio State University, and I thank the participants for their assistance.

CHAPTER 3

'All the Fits That's News to Print': Deinstitutionalisation and Anti-Psychiatric Movement Magazines in the United States, 1970–1986

Alexander Dunst

Introduction

This chapter analyses anti-psychiatric magazines that were at once the consequence of post-war deinstitutionalisation and participated in the further diversification of mental health care at a crucial time for US psychiatry. My discussion focuses on two closely related but ultimately separate social movements—radical therapy and mental patients' liberation—and the three publications that functioned as their public organs: *The Radical Therapist*, later renamed *Rough Times* and *State and Mind* (1970–76); *Issues in Radical Therapy* (1973–1986) and *Madness Network News* (1972–86). Initiated by disgruntled physicians and nurses, radical therapy rejected the existing mental health system for community care and constituted a small but influential element within the wider 1960s counterculture. Radical therapists sought social transformation and at times advocated outright

A. Dunst (✉)
University of Paderborn, Paderborn, Germany
e-mail: alexander.dunst@gmail.com

D. Kritsotaki et al. (eds.), *Deinstitutionalisation and After*,
Mental Health in Historical Perspective,
DOI 10.1007/978-3-319-45360-6_3

revolution, but also contributed to a popular self-help psychology that spelled a retreat from the political goals they espoused. Mental patients' liberation came into being in close cooperation with radical therapy but soon distanced itself from professionals to espouse a separatist politics inspired by Black Power and 1970s feminism. Focusing their efforts on communal self-assertion and patient rights, liberation activists engaged in protests and boycotts, initiated successful lawsuits and established patient-run alternatives to mainstream psychiatry.

Anti-psychiatric periodicals rarely feature in histories of mental illness but offer more than simply a neglected source of information. Edited by mental health workers and former patients, sent out to thousands of subscribers and sold in countercultural bookstores, they helped spread ideas and stimulated debates between professionals and non-professionals. *The Radical Therapist*, *Issues in Radical Therapy* and *Madness Network News* were the media through which protests were organised and where readers looked for like-minded people to start community centres, or shared their experiences as hospital patients. Spreading far beyond metropolitan centres, professional associations or single hospitals, these magazines established regional, national and increasingly international contacts and attracted new participants to the cause. In many ways, they *were* the movement. As one healthcare worker recalled years later: 'Once I got to (social work) school [...] I began to [...] follow old issues of the journal *The Radical Therapist*. I was intrigued and discovered I could do therapy.'[1] Most importantly, these publications acted as decentralised therapeutic spaces at a time when former patients found themselves housed in welfare hostels and care homes, or even lived on the streets facing economic insecurity and social stigma. The pages of these magazines represented a counterweight to isolation and oppression, and gave marginalised people an opportunity to share their own and learn from each other's experiences.

This chapter proceeds more or less chronologically, beginning with the radical therapy movement in section one, and then moving on to ex-patient activism in section two. The publications are discussed in the context of anti-psychiatric thought, specifically the writings of Thomas Szasz and R. D. Laing, and the establishment of national and international networks that connected scholars, activists and former psychiatric inmates on both sides of the Atlantic. Throughout, I pay attention to debates

[1] Cited in D. Wagner (1989) 'Radical Movements in the Social Services: A Theoretical Framework', *Social Service Review*, 63: 2, 273.

featured in these publications that discuss the downsizing of large psychiatric hospitals and their replacement by often private services favoured by neo-liberal reformers during Ronald Reagan's governorship in California and his two terms as president. While radical therapy ebbed away in the 1980s, ex-patient activism remains a viable movement to this day. Yet, the rise of liberal advocacy groups such as the National Alliance for the Mentally Ill (NAMI) in the early 1980s and increasing divisions among former patients means that the demise of *Madness Network News* in 1986 effectively marks the end of an era.[2]

Revolutionary Healing: The Radical Therapist and Issues in Radical Therapy

Radical therapy stood at the intersection of so many post-war developments that it may be described as a quintessentially 1960s phenomenon. Founded by disillusioned mental health workers, it resulted from an explosion in psychological services and contributed to their further diversification. Emerging in the final years of the decade and blossoming in the 1970s, radical therapy bore the imprint of a society which was splintering into numerous subcultures and causes, from feminism to more esoteric quests for happiness, and the further radicalisation of a left fringe. In an early issue, the editors of the *Radical Therapist* (*RT*) laid out a pluralistic understanding of what such therapy could amount to: organising a community against war, racism and environmental damage; awareness of the social rather than individual causes of psychic trouble; or opening a therapy centre for the young. The list concluded on a note that unified all of these under the roof of countercultural experience: 'Life-style is crucial.'[3] For the activist-journalists who published this and other dissident magazines, radical therapy was grounded in political engagement and sought fulfilment in everyday conduct.

Still, a marked tension existed between the two: the movement's radical wing, which sought to unify Marxism and psychology, berated the new therapy collectives and drop-in centres for their '"do-your-own-thing" ideology', accusing them of an abdication of duty.[4] For Marxist psychologists,

[2] NAMI is now called the National Alliance on Mental Illness. See section three for a comment on the implications of its name.

[3] Anon. (Jan.–Feb. 1971) 'On Practice', *The Radical Therapist*, 1–5, n.p.

[4] P. Brown (Fall 1977) 'Political Psychology', *Issues in Radical Therapy*, 20, 24.

political liberation trumped a liberated life-style—a priority that wasn't at all self-evident to those who sought healing in self-help. New Haven, Connecticut's, Number Nine, an alternative or 'counter-institution', as it called itself, was fairly typical. Named after the Beatles' song 'Revolution 9' and founded in late 1969, it employed 15 full-time staff and provided ten spaces for young people going through personal crises. As importantly in the context of the hippie movement, the centre came with its own psychedelic bus and rock band. Number Nine offered the usual mix of individual and group therapy, a drop-in centre and an emergency phone line. What brought all of these aspects together under one roof was a somewhat fuzzy sense of mission. As one staff member wrote: 'Our activities are unified through our growing [...] conception of ourselves as a community within a larger alternative culture, which is working for broad changes in the fabric of our society.'[5] In line with this emphasis on communal living, the house prided itself on its acceptance of difficult behaviour. Former patients often became members of staff, all of who lived together in a commune. Concrete goals were harder to come by: beyond the rejection of the status quo, the article limited its vision of an alternative society to the phrase 'growth through counseling'.[6]

Such vague conceptions of political change laid bare another contradiction: radical therapists defined their movement in fundamental opposition to the mental health care system but often provided alternative services instead of direct contestation. Rather than seeking alliances with sympathetic psychiatrists, nurses and ex-patients, they opened community centres and offered group therapy to other members of the counterculture. Frequently, these patients turned out to be as white and middle class as the therapists who helped them. One essay in *Issues in Radical Therapy* cautioned that the movement was limited to major cities and liberal university towns, 'and as of yet has had no impact on any third world community'.[7] Similar problems plagued most 1960s liberation movements, from radical student activism to feminism. However, this class base helps explain a crucial oversight across the pages of these journals. With few exceptions, the gathering storm of a genetic and biochemical psychiatry emits no response

[5] D. Jaffe (Jan.–Feb. 1971) 'Number Nine: Creating a Counter-Institution', *The Radical Therapist*, 1–5, 10.

[6] Ibid., 11.

[7] J. Aronson and E. Klauber (Spring 1977) 'Counter-Hegemony and Radical Psychiatry', *Issues in Radical Therapy*, 17, 3.

in them. Countercultural therapists and their clients not only sought distance from mainstream institutions and psychiatric research, they were rarely at the receiving end of the drug treatments that most patients with serious illnesses, or from poorer backgrounds, received. We may also attribute this curious absence to the movement's conviction in the social causes of mental illness and the necessity of political change to combat it. In other words: why engage with what you already know to be false?

Radical therapy therefore presents a case study of the gap that opened up between self-consciously political understandings of madness in the counterculture and the turn towards medical models in psychiatry. To an extent, this split persists even today: between medical practitioners and many of their colleagues in social work, nursing and psychology, and also between humanities and social science scholars and researchers in the biological sciences. Radical therapy journals were central to this development. Psychiatrists Michael Glenn and Michael Galen had taken part in the APA's annual meeting in Miami in 1969, during which its liberal caucus adopted the moniker radical for the first time. Frustrated with the APA's resistance to progressive demands, Glenn and Galen returned to the Air Force base at Minot, North Dakota, where they were stationed, to start *RT* in the following spring. Claude Steiner, the movement's chief theorist and another member of the radical caucus, began publishing *Issues in Radical Therapy* (*IRT*) three years later after opening his own alternative centre in Berkeley.[8] For these psychiatrists, the break with the APA also led to a gradual distancing from their own professional identity—which wasn't always a smooth process.

RT characterised itself as a '"professional" journal' in the founding issue, but Glenn's opening manifesto also called for the demystification and de-professionalisation of psychotherapy.[9] What exactly this process entailed constituted a central debate of early issues. Some contributors advocated a process of sharing skills to democratise therapy. Others extended Szasz's famous argument about mental illness to declare skills a myth.[10] Much of the frustration stemmed from what radical therapists felt to be the irrelevance of a medical education to their daily work. In Steiner's training centre, more

[8] L. Richert (2013) '"Therapy Means Change, not Peanut Butter": American Radical Psychiatry, 1968–1975', *Social History of Medicine*, 27:1, 112–13.

[9] M. Glenn (Apr.–May 1970) 'Manifesto', *RT*, 1:1, 2.

[10] T. Szasz (1974) *The Myth of Mental Illness: Foundations of a Theory of Personal Conduct*, rev. edn (New York: Harper & Row).

conventional instruction was replaced with 12 months of collective learning that rejected psychiatric terminology and individual sessions for 'soul healing' in a group setting.[11] This alternative institution was titled the Radical Psychiatry Center, but as Steiner acknowledged, the reference to psychiatry was somewhat of a misnomer and meant primarily as a provocation. Despite the rhetorical similarities, what emerged from the early debates about professionalism was distinct from the writings of well-known critics like Szasz, Erving Goffman, Michel Foucault and Laing. Neither Szasz nor Laing, the two psychiatrists among them, showed much interest in making therapeutic techniques available beyond their own profession. In contrast, radical therapy's emphasis on skill-sharing and non-professional healing allowed it to become a movement that was anchored around two journals but extended to community teaching, drop-in centres and regular conferences.

As the movement's primary means of communication, *RT* and *IRT* showed many parallels with the 1960s dissident press.[12] Made possible by the availability of cheap photo-offset printing, the era saw a boom in underground publications that made up for amateurish layout and design with unabashed idealism. The editors of *RT* and *IRT* saw themselves as activists first, journalists second—priorities reflected in the uneven quality of writing and inconsistent spelling to be found in their journals. Radical therapy publications also shared a chronic lack of money and establishment hostility with their more distant forerunners in antebellum abolitionism and the nineteenth-century labour movement. All these publications linked and shored up communities, supported and in many cases initiated grassroots action, and opened their pages to the communities they served. *RT* and *IRT* counted prominent reformers, scholars and novelists as their authors, among them Barbara Ehrenreich, Deirdre English, Todd Gitlin and Marge Piercy. Yet their most notable characteristic was a fundamentally democratic ethos: produced by a collective of activists, subscriptions were offered free to prisoners, mental patients and anyone who suffered from a 'real lack of money'.[13] Articles, letters and poems by current and former psychiatric inmates filled their pages. Others contacted the jour-

[11] C. Steiner (June–July 1970) 'Radical Psychiatry – Manifesto', *RT*, 1:2, 12.

[12] For an overview of underground publications see R. Streitmatter (2001) *Voices of Revolution: The Dissident Press in America* (New York: Columbia University Press); and for a more detailed study of the 1960s, J. McMillian (2011) *Smoking Typewriters: The Sixties Underground Press and the Rise of Alternative Media in America* (New York: Oxford University Press).

[13] *RT*, 1:6 (Feb.–Mar. 1971), 1.

nals when they wanted to start their own therapy centre. As reader C. A. Fairfield, describing himself as a former psychiatric aide in Los Angeles, wrote:

> I want to establish a commune in Northern California whose 'thing' it will be to provide the setting for schizophrenics to ride out their trips sans shock sans drugs sans shrinks (except a hip shrink or whathaveyou) and sans silly ass psychiatrisms—do you know of anyone already doing the Laing-like community in a rustic setting?[14]

This countercultural outlook continued in the visual makeup of the magazines. In keeping with other publications, *RT* and *IRT* freely reprinted articles and illustrations that were shared via the Underground Press Syndicate and Liberation News Service. Cartoons that expounded Mao quotes alternated with photographs from political rallies, woodcuts by Frans Masereel and comic strips by Robert Crumb. Drawings of the raised and clenched fist—made popular by Black Power—could be found next to advertisements for Atlanta's 'Liberation Records' and the 'Gay Liberation' magazine.

Radical therapy publications also distinguished themselves from the wider underground press. Serving a tight-knit community on the edges of a huge movement its publications reached a far smaller audience, which probably never exceeded *RT*'s print run of 5,000 copies and a roughly equal number for *IRT*.[15] In contrast to magazines like the Los Angeles *Free Press* or the *Chicago Reader*, which could count on a large urban readership, theirs was a national and sometimes international audience. While they could be described as social movement media, the two journals also resembled special interest publications and academic journals in their thematic focus and specialised vocabulary. Already *RT*'s founding editorial, its title asking the question 'How Revolutionary Is a Journal?', confronted this hybrid identity.[16] It admitted that such publications, aimed at a highly literate audience, ran the danger of being elitist and ineffective. In self-defence the editors argued that a journal served to clarify issues and rally support, bringing its

[14] C. A. Fairfield (Oct.–Nov. 1970) 'Letter', *RT*, 1:4, 2.

[15] The relative obscurity of the two magazines also meant that they escaped concerted efforts by local police and the Federal Bureau of Investigation to harass, infiltrate and undermine the underground press. See Streitmatter, *Voices of Revolution* and McMillian, *Smoking Typewriters* for documentation of these frequently illegal activities.

[16] The Editors (Apr.–May 1970) 'How Revolutionary Is a Journal?', *RT*, 1:1, n.p.

readers together for social action. This rhetorical commitment was soon followed by the founding of a grassroots institution called Open in Minot. After town pressure the local community health centre withdrew its offer to provide space in an unused building, but Open soon offered a training group for aspiring therapists, welcomed a women's liberation and GI group, and provided drop-in services for college and high school students.[17]

In the journal's second year, an ideological struggle erupted. An issue dedicated to writings by members of Berkeley's Radical Psychiatry Center was met with an editorial that questioned the Californians' involvement in wider struggles and accused them of a 'middle-class emphasis on groovy fun and individual solutions'.[18] Coinciding with a move from North Dakota to Somerville, Massachusetts, a suburb of Boston, these debates led to a name change and the birth of *IRT*. After the final issue of 1971 had critiqued the journal's title *Radical Therapist* (radical for its trendy over-use, therapy for its professional connotation), it began appearing as *Rough Times* instead. The motivations for this switch were made abundantly clear in June 1972: 'We participate in the world-wide revolutionary force, not within the confines of "radicals in the professions." We want [...] total revolution, not just reforms.'[19] The shift in editorial policy showed in a special section on 'Vietnam & Psychology', articles on mental patients' liberation and the German Socialist Patient Collective, located in Heidelberg and equally committed to revolution. Increasingly, *Rough Times* looked beyond America's borders to guide the movement's efforts. Wilhelm Reich mixed with Frantz Fanon, and the ever-present Szasz was discussed next to the Italian reformer Franco Basaglia. Reich, in particular, highlighted the contradictions between this eclectic blend of writers and their uptake by alternative therapists. His combination of Marxist and psychoanalytic credentials endeared Reich to radical psychologists but the emphasis on sexual liberation played into the hands of a narcissistic retreat into private life, whether that took the shape of a commune or more traditional arrangements. As the feminist writers Ehrenreich and English warned in an astute analysis a couple of years later, the supposed radicalism of sexual liberation fed a commercial pop psychology that sapped the flame of left-wing politics.[20]

[17] Anon. (Jan.–Feb. 1971) 'RT in Practice', *RT*, 1:1, 17.

[18] The Editors (Oct. 1972) 'Critique', *RT*, 2:3, 31.

[19] Anon. (June 1972) 'Editorial', *Rough Times*, 2:7, 1.

[20] D. English and B. Ehrenreich (Summer 1974) 'Politics of Sexual Freedom', *IRT*, 2:3, 15–17.

In the meantime, *IRT* provided a new forum for the radical psychiatry group around Steiner. Emerging after the fallout with *RT*'s editors, the West Coast journal featured higher production values, as well as longer and more academic articles, but understood itself as part of the same movement. The founding editorial referred to *IRT* as a practical journal, emphasised community organising and vowed to resist co-optation by liberal psychiatrists and a hip, depoliticised counterculture. In 1976, the editorial collective together with local organisers launched the 'Radical Psychiatry Summer Institute'. The same year, the 'Midwest Radical Therapy Conference' attracted 250 participants to a campsite near Iowa City. Workshops during the four-day event included 'feminism for men', 'visual techniques in therapy' and 'herbal and psychic healing'.[21] Regional initiatives coexisted with growing internationalisation: British psychoanalyst Susi Orbach joined as a contributing editor; the summer institutes brought the American movement in touch with German, Canadian and Swedish colleagues; and articles focused on Basaglia's work in northern Italy, Gilles Deleuze and Félix Guattari's *The Anti-Oedipus*, as well as post-revolutionary psychiatry in Nicaragua.

Therapeutically, these examples remained in the shadow of Steiner's own approach, which he distilled from Thomas Harris' hugely successful self-help bible *I'm OK, You're OK*. Eric Berne, the founder of transactional analysis, had rejected diagnostic concepts for a therapeutic encounter that focused on social interaction. Despite acknowledging Berne's apolitical liberalism, Steiner based his therapy on Berne's and Harris' work and put the blame for mental illness on families. Two basic premises guided Steiner: the first was a chimerical belief in absolute harmony, the second saw alienation at the root of all psychiatric conditions. At this point, the psychoanalytic spotlight on family dynamics combined uneasily with an understanding of alienation as political oppression. As a consequence, Steiner declared that healing could not be based on individual insight but had to evolve out of a group fighting for its liberation. Precisely how this journey from awareness to action should be accomplished remained a bone of contention throughout *IRT* and radical therapy's existence. The Szaszian elements in Steiner's worldview proved less controversial. The rejection of medical terminology perhaps owed as much to Szasz as to Berne, as did the caveat that excluded diseases with physiological causes when mental illness was declared a myth. These debts were thrown into

[21] Anon. (Summer 1976) 'Movement News', *IRT*, 4:3, 29.

stark relief in an interview Steiner conducted with Laing in the latter's home in London in 1975. By then, Laing had repudiated his left-wing rhetoric of the late 1960s and was living contentedly, as he claimed, in a nuclear family of his own. After Laing's curt dismissal of communes—'they don't seem to be particularly better', he observed—the conversation turned to madness. Steiner's insistence that madness, like mental illness, represented merely a form of emotional upset was met with derision: 'Well, some people are pretty crazy', Laing replied.[22]

The contradiction between healing and change became only more acute as the counterculture's verve evaporated in the late 1970s. After swinging back from an extreme left position to self-help, *RT* reported financial problems and, after a final name change to *State and Mind*, bowed out in 1976. Now the sole publication of the movement, *IRT* showed greater longevity and was published in the Bay Area until 1983, when editorial responsibility shifted to the Illinois Radical Therapy Collective in Springfield for another 3-year period. The journal became more streamlined, its briskly written editorials approximating the mainstream press. With Reagan's election to the presidency, the ascent of the right began to worry contributors but the changing times also showed within the movement. Rhetorically, *IRT* remained steadfast in its commitment to socialism and collective healing. Some articles and movement activities struck a different note. Reporting from a 'Radical Therapy Theory Conference', held in February 1984 in San Francisco, the author complained about the time wasted by the democratic process of agenda setting and advocated individual accountability to replace the '"tyranny of structurelessness"'.[23] The meeting had been organised by an alliance of ex-patients, therapists and academics but the article's managerial perspective was a far cry from the spirit of the 1960s, as was the focus on theory and the move to limit participation to invited speakers.

The gradual absorption of radical therapy into the mainstream was unmistakable. Similar in some ways to the fate of ex-patient liberation, this outcome recalled one of the most piercing assessments of the movement. Written by Mona Field, of the Radical Therapy Training Collective in Los Angeles, the article confronted movement members with a narrower but more painful failure.[24] By 1977, when Field's short essay was published,

[22] Steiner and Meighan 'R. D. Laing: An Interview', *IRT*, 6 and 8.

[23] P. Parsons (Spring 1984) 'Radical Therapy Theory Conference', *IRT*, 11:2, 30.

[24] M. Field (Summer 1977) 'Left Turn at Echo Park', *IRT*, 19, 7.

revolutionary change in the US had receded even for die-hard believers. At a time when the dissident press was fading, dealt a final blow by the end of the Vietnam War, radical therapists continued to publish their own journal, organise conferences and run community centres. This alone counted as a remarkable success. Yet the overall picture appeared bleak: Field's community activism in a traditionally left-wing neighbourhood had taught her that far from reaching the masses of ordinary Americans, or mounting an effective challenge to institutional psychiatry, radical therapists hadn't even convinced fellow liberals. Yes, many people viewed conventional psychotherapy as oppressive, an argument that writers like Steiner, Glenn and Galen had made from the start. Neither, however, did they recognise radical versions of it as a political tool. To them mental illness remained a personal, not a political, problem.

Madness Network News Between Liberty and Neo-liberalism

Despite considerable overlap, radical therapy and ex-patient activism constituted two separate movements at their inception, and drifted apart further as they unfolded. This division appears by no means inevitable. *Madness Network News* (*MNN*), the voice of mental patients' liberation from 1972 to 1986, was the brainchild of left-wing mental health workers and former patients in the Bay Area of California. Boston's Mental Patients' Liberation Front, one of the most vociferous advocacy groups to emerge, held its first meetings in the offices of *RT*, which had only recently moved from North Dakota to Massachusetts. Similarly, patient activists such as Judi Chamberlin or Leonard Roy Frank, *MNN*'s long-time editor, contributed to *RT* and *IRT*. Regular conferences, launched in Detroit in 1973, provided a shared platform until separatism took hold.

Radical therapy and mental patients' liberation came into being as a consequence of psychiatry's explosive diversification in the post-war era, but they constituted different elements within it and their proponents took their place on opposite sides. Radical therapy evolved out of the political activism of left-wing psychiatrists and remained, despite calls to de-professionalise, a mainly middle-class phenomenon within the countercultural bohemia of the 1960s. Mental patients' liberation took root in a very different milieu. Patient publications had existed before in the US, but *MNN* spoke for a movement because a majority of psychiatric patients came to live outside large hospital complexes, where they had once been housed. Released more quickly and at an earlier age, many found that an existence outside psychiatric institutions was fundamentally different but not necessarily easier as they were forced to live on meagre benefits, struggled with homelessness, unemployment or menial jobs, and faced social stigma and often fragile health. In this situation, patients depended on peer-support like never before. For those who were still in hospital care, too ill, or lived far from alternative centres, *MNN* provided a therapeutic space in which the community could share its experiences and discuss solutions in a sympathetic environment. One recurrent problem addressed in the magazine was the reactions of new acquaintances or potential employers to a history of institutionalisation. *MNN*'s editors encouraged readers to come out, often with success. A reader who gave her name as Virginia Davis wrote to thank the magazine for helping her to make this important step: 'It's becoming easier and easier to say, "I've been crazy".'[25]

As the 1960s drew to a close, self-help groups spread rapidly across the country, starting with Portland's Insane Liberation Front in 1970, the Mental Patients' Liberation Project in New York City a year later, and the Network Against Psychiatric Assault (NAPA) in San Francisco in 1972, to name just a few. Activists in these organisations often came from working-class backgrounds and found collective living to be an economic necessity rather than an aspirational lifestyle. Their foremost goals were the establishment of support networks and patient rights: the freedom to choose and—above all—to refuse treatment. Upon reading *MNN*, it quickly becomes evident how low revolution and countercultural experience rank among the priorities of its contributors. It would be a misunderstanding, however, to dismiss the movement's political nature. The insistence on freedom from treatment added direct contestation to the construction of alternative institutions—from demonstrations to boycotts and lawsuits. Radical therapy might have had loftier ambitions, but for more than a decade mental patients' liberation was built around a core of fundamental opposition.

This section focuses on the dominant strand of ex-patient activism of the 1970s and early 1980s. This era coincides with the meetings that brought activists and sympathisers together at the yearly 'Conferences on Human Rights and Against Psychiatric Oppression', and the existence of

[25] V. Davis (June 1974) 'Letter to the Editor', *Madness Network News*, 2:3, 23.

its major public platform. Sold for 50 cents (later 75 cents), *MNN* boasted a print run of 4,000 copies, roughly the same as *RT* and *IRT*. Under the title, the header riffed on no less than the *New York Times* to proclaim itself mad America's newsletter of record: 'All the Fits That's News to Print'. Where radical therapy publications leaned towards professional journals in their terminology, *MNN* shows many historical parallels with black, gay and women's liberation magazines like *Black Panther* and *Come Out!*, first published in late 1969.[26] Advertisements for *MNN* printed in other countercultural newspapers referred knowingly to the kinship between these movements: an illustration showed a wild-haired man, dressed in flowing garments, mouth opened wide as if shouting, his right fist raised high. These publications sprang up as the civil rights and student movements split into identity-based creeds and countercultural publications began to address more closely-knit communities. Begun as a community newsletter for the Bay Area, *MNN* soon attracted an international readership and contributors. Some of these were disillusioned psychiatrists, but from the beginning the main emphasis lay on the self-assertion of current and former patients. Aiming to protect the 'rights and dignity of those people labeled crazy', the editors called for patients who sent in letters, poems, illustrations and essays to give their real names in order to 'break the myth of mental illness'.[27]

An early contribution of this kind was 'Shock', a short article written by Cyril Kolotronis, who gave his address as Western State Hospital in Lakewood, Washington. The author detailed his ECT treatment, mourning the memory loss that left him unable to do math and which made him much worse, he added, at playing Bridge and Pinochle. Kolotronis wasn't content merely to describe his experiences in neutral fashion, drawing on popular Cold War imagery to denounce psychiatry. Electroshock constituted a war crime, committed by a 'foreign doctor, who went to the same school as Fidel Castro' and recalled the medical experiments conducted in Auschwitz.[28] 'Can this be America?', Kolotronis asked in closing. The essay wildly mixed its references to communism and national socialism, but the understanding of psychiatry as a totalitarian institution engaging in torture was fairly standard in the pages of *MNN*. Like the radical therapy press, *MNN*'s contributors occasionally dabbled in Reich and Fanon, but

[26] Streitmatter, *Voices of Revolution*, pp. 223–25 and 243.
[27] Anon. (1973), *MNN*, 1:5, 2.
[28] C. Kolotronis (1973) 'Shock', *Madness Network News*, 1:5, 9.

the magazine's ideological outlook was again dominated by Szasz. Szasz's denial of mental illness, rephrased as non-medical problems in living, led to *MNN*'s insistence that treatment should only ever be on a voluntary outpatient basis, even if the alternatives were the 'mental patient ghettos' that several contributors reported on—the private care homes and welfare hostels that were springing up in increasing numbers in city centres and suburbs.[29] Rarely was this libertarian world view questioned. In one such debate, a member of the self-consciously Marxist Mental Patients' Liberation Front questioned calls for individual responsibility that had arisen in the context of violence attributed to released mental patients. Instead, the article called for a revolutionary overthrow of oppressive institutions. The reaction by *MNN*'s editorial board was swift: 'As far as responsibility goes. What's wrong with that? You quote Karl Marx, then reach a conclusion that just doesn't follow.'[30] Szasz himself was fully aware of his prominent position in the movement and expressed his admiration for *MNN*, sending a cheque for a 2-year subscription with his letter.

The insistence on individual responsibility formed a central plank of ex-patient activism. Not only did this sit rather uneasily with Szasz's occasional remarks about 'lazy or lethargic, sick or stupid' mental patients, it also begged the question what distinguished them from other downtrodden people with so-called problems in living.[31] Such reflections could have opened the way for working-class alliances but their pursuit was cut off by identity politics and a purely negative conception of liberty as freedom from treatment. Self-assertion also posed problems. Celebrations of insanity, from a 'Festival of Creative Psychosis' to Laingian workshops on madness as breakthrough, foreshadowed the Mad Pride movement but contradicted the reduction of mental illness to social difficulties.[32] These strategic deficits notwithstanding, the yearly conferences and *MNN* became a rallying ground for legal and public activism. In 1975, a lawsuit filed by seven inmates of Boston State Hospital, who had been members of the local liberation group, established a limited right to refuse drug treatment in Massachusetts, with similar reforms following in New York and

[29] B. Harris (Fall 1977) 'Testimony by Ex-Inmate Activists', *MNN*, 4:4, 32.

[30] S. Soldz (Feb. 1974) 'Letter to the Editor', *MNN*, 2:2, 21; and S. Hirsch (Feb. 1974) 'Dialogue', *MNN*, 2:2, 22.

[31] Szasz, *The Myth of Mental Illness*, pp. 164 and 175.

[32] Crossley gives a brief overview of the evolution of the British Mad Pride movement in N. Crossley (2006) *Contesting Psychiatry: Social Movements in Mental Health* (New York: Routledge), pp. 205–207.

California.[33] In 1976, a month-long sit-in at the office of Governor Jerry Brown led to the establishment of advocacy groups in California's counties. Celebrating this success, *MNN*'s October issue printed a photograph of several women sitting on a couch in the governor's office and holding up placards that read: 'Mad Women Fight Back' and 'Bet Your Ass We're Paranoid'.[34] In a later campaign, movement activists succeeded in adding a question on banning ECT treatment to the city election ballot in Los Angeles. Soon after it was passed with more than 60 per cent of votes in 1982, the ban was lifted again because of a lawsuit by psychiatrists.[35]

These partial successes might have contributed to a more sustained legacy had it not been for another far-reaching error. After complaining about the elitism of their professional allies at the 1975 conference, patients started holding closed meetings. Chamberlin fondly recalled: 'for the rest of the Conference, most of the ex-patients could be found behind a door with a very clear "Keep Out" sign for anyone who hadn't done time in a mental institution'.[36] Activist groups in Boston and New York became limited to former patients, and in 1976 *MNN* followed suit after what Chamberlin described unceremoniously as a coup. Taking inspiration from Black Power and radical feminists, separatism was a reaction to very real differences in class and power. It was also rooted in contrary political goals: radical psychiatrists, even where they defended patients against physical and drug treatment, advocated therapy as an emancipatory tool and lobbied for its expansion. Liberation activists, in contrast, viewed any attempt to cure them sceptically and often called for the complete abolition of therapeutic intervention. Given this opposition, a lasting alliance between the two groups would have been difficult to sustain. Yet separatism, pursued by such a small and powerless group with few natural allies, further undermined the movement.

These weaknesses took their toll once the political climate turned against radical movements. In the course of 1975, a new self-definition crept into *MNN*: the patient became a consumer. The term would become more frequent over the following years, but for the time being this tendency co-existed with more radical self-assertions. At the seventh annual conference

[33] J. Chamberlin (1990) 'The Ex-Patients' Movement: Where We've Been and Where We're Going', *The Journal of Mind and Behavior*, special issue 11:3/4, *Challenging the Therapeutic State: Critical Perspectives on Psychiatry and the Mental Health System*, 333.

[34] See back cover of *MNN*, 4:1 (Oct. 1976).

[35] J. Chamberlin (1985) 'The Ex-Inmates' Movement Today', *IRT*, 11:4, 15.

[36] J. Chamberlin (Oct. 1976) 'Organising? or Disorganising?', *MNN*, 3:3, 4.

in 1979, a Dutch activist became the first non-American speaker to feature at the event following a visit from European ex-patients. *MNN* increasingly found a wider echo. As early as 1972, the author Ken Kesey had written to the magazine to thank them for updating him on recent developments while he was working on the screenplay for Milos Forman's adaptation of his novel *One Flew Over the Cuckoo's Nest*. In 1977, Guattari visited liberation groups in New York, sharing his address in Paris with *MNN*; around the same time, a French reader of *MNN* reported that she had passed a few issues on to Foucault.[37] Special sections on the European anti-psychiatry movement were followed by articles on Germany's Socialist Patient Collective and an essay by Basaglia. Given the dearth of financial resources in the movement, these grassroots exchanges were quite remarkable.

In the meantime, mental patients' liberation faced growing competition on its home soil. Founded in 1979, the National Alliance for the Mentally Ill (NAMI) would eventually develop into the largest advocacy group for psychiatric patients in the country. The name was telling in itself: where ex-patients had asserted their right to speak in their own voice, the new group consisted of parents and relatives who spoke *for* them. These families, often more middle-class and respectable, as *MNN* noted, began to build the alliances with existing institutions and sympathetic professionals that liberation activists largely had shunned. Hand in hand went their enthusiastic adoption of biological models of mental illness: support for genetic and biochemical explanations aligned NAMI with the new psychiatric orthodoxy, but also absolved parents from accusations that they were responsible for their children's illness.

Arguably, the libertarian underpinnings of mental patients' liberation laid the groundwork for NAMI's later emphasis on consumer choice and feedback. What allowed it to grow in influence so rapidly, however, were market-driven reforms. During Reagan's tenure as governor, California's healthcare system became a testing ground for the neo-liberal ideas he would unleash nationwide during his presidency. Under the cover of strengthening individual rights, the so-called Lanterman–Petris–Short Act, coming into full effect in 1972, led to a further drop in patient numbers. *MNN* explicitly defended the law from criticism, but the reform played into the hands of organisations like NAMI.[38] When the mainstream

[37] L. Morrison (2005) *Talking Back To Psychiatry: The Psychiatric Consumer/Survivor/Ex-Patient Movement* (New York: Routledge), p. 72.

[38] Anon. (Dec. 1973) 'Notes from LAMP … about psychiatry and the law', *MNN*, 2:1, 8–10.

media blamed a number of sensational murders on released patients, many of whom now found themselves crammed into hostels or homeless, moderate self-help groups were at hand to alleviate the effects of understaffed and underfunded psychiatric wards, or simply to provide a fig leaf for politicians eager to deflect from the consequences of their actions. The founding of their own National Alliance of Mental Patients by liberation activists in 1985 came too late—at least for their long-standing movement organ. *MNN*'s final issue announced the dissolution of the editorial collective, closing with a bleak outlook into the future:

> A real lack of communication and trust in recent years among people has been another major cause [...] [of] co-opting the movement [...] Very few anti-psychiatry projects remain. Those that do tend to be isolated. It remains to be seen whether or not a visible, organized anti-psychiatry movement will continue to exist.[39]

Things did not turn out to be quite as bleak. Still, the collapse of *MNN* initiated a long, dry spell for radical patient activism. The government funding that became available for self-help initiatives in the 1980s flowed elsewhere. In another symbolic defeat the yearly movement conferences were newly titled 'Mental Health Alternatives'—a name change that signalled the end of principled opposition. Nevertheless, *MNN*'s unique importance lies in the way in which it documents a creative response to institutional psychiatry that arose out of the social and cultural maelstrom of the sixties, and the new lives that former patients began to live outside large state hospitals.

Conclusion

The anti-psychiatric magazines I have discussed in this chapter came into being during a drastic reduction in the number of mental patients in the United States. Yet this development alone, which had begun two decades earlier, does not explain their success. Radical therapy emerged in the tailwinds of widespread upheaval in the latter half of the 1960s and appropriated the political and media tactics of social movements led by students, civil rights, gay and feminist activists. Equally important proved prominent critics of psychiatry such as Goffman, Laing and Szasz. Their writings laid

[39] *MNN* 8:3 (1986), 2; cited in Morrison, *Talking Back to Psychiatry*, p. 82.

the foundations for radical therapy but its aims were more specific. Calls to de-professionalise—to reject expertise and doctor–patient hierarchies for group therapy and non-medical healing—constituted a direct challenge to psychiatric orthodoxy. In so doing, progressive therapists contributed to deinstitutionalisation, but the term captures only one side of this dynamic. Radical therapy did not advocate the abolition of psychiatric help. In fact, it argued for its expansion and the replacement of existing services with democratic institutions in the community. If this reform programme contributed to the closure of large state-run hospitals, this success was due less to the power of critique than to radical therapists establishing their own institutions—from therapy centres to self-help groups and movement magazines. The latter were far from immaterial: they developed and disseminated new ideas and movement tactics, helped organise events, found new groups and connected therapists to patients. Without this communicative network, establishing a vibrant national and international movement might have proven impossible for a small group of radical mental health workers.

A similar dynamic may be seen at work in mental patients' liberation, which initially overlapped with radical therapy but then diverged from its path. Like radical therapists, ex-patient activists assembled an identity from their shared experience of psychiatric institutions and sought alternatives to them. In organising for change, they by necessity founded new institutions: publications like *Madness Network News*, halfway houses and other meeting places, as well as the yearly conferences that brought together activists from across America and Europe. As Crossley has argued for Britain, radical ex-patients eventually fell victim to their own limited success.[40] As time went by, the opinions and perspectives of ex-patients were increasingly sought out by institutions and government bodies. Judi Chamberlin, for instance, served on Jimmy Carter's President's Commission on Mental Health in the late 1970s.[41] If she ultimately rejected the compromise that official recognition demanded, others were more willing to don the mantle of mental health consumers and initiate a new phase in the history of psychiatric reform.

[40] Crossley (2006) *Contesting Psychiatry*, 191.
[41] Chamberlin (1985) 'The Ex-Inmates' Movement Today'.

CHAPTER 4

After the Asylum in Canada: Surviving Deinstitutionalisation and Revising History

Megan Davies, Erika Dyck, Leslie Baker, Lanny Beckman, Geertje Boschma, Chris Dooley, Kathleen Kendall, Eugène LeBlanc, Robert Menzies, Marina Morrow, Diane Purvey, Nérée St-Amand, Marie-Claude Thifault, Jayne Melville Whyte, and Victor Willis

Introduction

Psychiatric deinstitutionalisation began in Canada in earnest during the 1960s and continues today.[1] The downsizing and closure of custodial

[1] M. Morrow and B. Jamer (2008) 'Making Meaning in a "Post-Institutional" Age: Reflections on the Experience of (De)Institutionalisation', *International Journal of Psychosocial Rehabilitation*, 12:2, 1–13; M. Morrow, A. Pederson, J. Smith, V. Josewski, B. Jamer and L. Battersby (2010) *Relocating Mental Health Care in British Columbia:*

M. Davies (✉)
York University, Toronto, Canada
Kwantlen Polytechnic University, Surrey, Canada
University of Ottawa, Ottawa, Canada
Independent Author, Saskatchewan, Canada
Parkdale Activity–Recreation Centre (PARC), Toronto, Canada
e-mail: daviesmj@yorku.ca

D. Kritsotaki et al. (eds.), *Deinstitutionalisation and After*, Mental Health in Historical Perspective,
DOI 10.1007/978-3-319-45360-6_4

mental hospitals did not occur uniformly across the country, and regional variations in government, healthcare staff and community care policies profoundly shaped the process. The Saskatchewan Mental Hospital at Weyburn, the last asylum built in the Victorian style in the British Commonwealth, was the first to shut its doors, which it did dramatically in 1963.[2] Others closed in stages, emptying wings and transitioning into

Riverview Hospital Redevelopment, Regionalisation and Gender in Psychiatric and Social Care (Vancouver: Centre for the Study of Gender, Social Inequities and Mental Health).

[2] P. Sealy and P. Whitehead (2004) 'Forty Years of Deinstitutionalisation of Psychiatric Services in Canada: An Empirical Assessment', *Canadian Journal of Psychiatry*, 49:4, 249–50.

E. Dyck • L. Baker
Independent Scholar, Halifax, Nova Scotia, Canada
e-mail: erika.dyck@usask.ca; lbaker.usask@gmail.com

L. Beckman
Vancouver, Canada

G. Boschma
University of British Columbia, Vancouver, Canada
e-mail: geertje.boschma@nursing.ubc.ca

C. Dooley
University of Winnipeg, Winnipeg, Canada
e-mail: Chris_Dooley@MTS.net

K. Kendall
University of Southampton, Southampton, UK
e-mail: k.a.kendall@soton.ac.uk

E. LeBlanc
Groupe de support émotionnel inc, Moncton,, Canada

R. Menzies
Simon Fraser University, Burnaby, Canada
e-mail: menzies@sfu.ca

M. Morrow
York University, Toronto, Canada
e-mail: mmmorrow@yorku.ca

D. Purvey
Kwantlen Polytechnic University, Surrey, Canada
e-mail: diane.purvey@kpu.ca

outpatient care facilities or, as was the case in Alberta, repurposing the buildings for brain injured patients requiring shorter-term stays.[3] Some facilities remained open with a reduced patient population and abandoned sections of the hospital that no longer conformed to the standards for privacy or health and safety regulations. Eastern Canada provinces like Nova Scotia had not subscribed to large-scale custodial institutions in the first place, and while deinstitutionalisation from cottage-style facilities also occurred, the pace and impact of that change was profoundly different for staff, communities and ex-patients. Several Ontario-based institutions centralised their services, closing some and enlarging others. British Columbia's iconic Riverview mental hospital continued to exist partially until 2012, looming large in cultural memory, as did many of these other monuments to what soon became a bygone era of psychiatric care. This regional variation in service delivery has in part characterised deinstitutionalisation in Canada, and also helps to underscore how patients from place to place may have encountered very different circumstances as they moved out of institutional care.

In 2004, Canadian psychiatrists Patricia Sealy and Paul Whitehead assessed deinstitutionalisation over a 40-year period beginning in the 1960s. Arguing that the Canadian evolution challenged conceptualisations of deinstitutionalisation, Sealy and Whitehead suggested expanding the definition from a simple movement out of custodial care to a wider understanding that included a broader and more varied set of experiences. At its core, a weakened dependence on large-scale hospitalisation persisted, but the process of reorganising the system also introduced new features of transinstitutionalisation, or the movement through a new set

[3] See for example, the Ponoka Mental Hospital in Alberta, which changed its focus from long-term psychiatric care to acquired brain injury treatments.

N. St-Amand • M.-C. Thifault
University of Ottawa, Ottawa, Canada
e-mail: nstaman@uottawa.ca; mthifaul@uottawa.ca

J.M. Whyte
Independent Author, Saskatchewan, Canada
e-mail: jayne.whyte@sasktel.net

V. Willis
Parkdale Activity–Recreation Centre (PARC), Toronto, Canada

of institutions, whether penitentiaries, nursing homes, hospitals and/ or new psychiatric facilities. They found a distinctive rise in the number of short-stay admissions in general hospitals, alongside a proliferation of community-based outpatient services. Sealy and Whitehead concluded that deinstitutionalisation involved fundamental changes in service delivery and staffing, which ricocheted through the regions in uneven ways, creating new state expenditures and putting pressure on traditional systems to absorb the impact of these decisions.[4]

Situating deinstitutionalisation within a broader social, cultural and medical history, we seek to understand how deinstitutionalisation has been experienced directly. We examine the products of a new generation of research on deinstitutionalisation in Canada, one anchored in community-based partnerships and emphasising the perspectives of frontline consumers and staff in their appraisals of what deinstitutionalisation means and how we should memorialise the asylum. This analysis stems from a research project funded by a Canadian Institutes for Health Research (CIHR) grant to examine the history of deinstitutionalisation in Canada. The products of that study are hosted by the existing History of Madness/ Histoire de la folie website (http://historyofmadness.ca/). Co-directed by historians Megan Davies and Erika Dyck, the Après l'Asile/After the Asylum projects involved a multidisciplinary, multinational team of academic researchers from across Canada, the United States and Britain.[5] This chapter examines the process of establishing this community–university network and how those relationships contributed to enriching our historical appreciation for the contours of deinstitutionalisation.

History of Deinstitutionalisation in Canada

The history of deinstitutionalisation in Canada, as was the case in other locations, was intertwined with contemporary sociocultural transitions, political restructuring and economic conservatism. Deinstitutionalisation in the most literal sense involved the massive depopulation of mental hospitals across the country. But the motivations for this evolution were a complex blend of new therapeutics, professional change and fiscal planning, and changes wrought by deinstitutionalisation were far-reaching in

[4] Sealy and Whitehead, 'Forty Years of Deinstitutionalisation', 249–50.

[5] The original project name was *Open Doors/Closed Ranks: Locating Mental Health After the Asylum*.

their impact on the Canadian economy and on human rights discourses, particularly those that hinged upon questions of ability, disability, health and illness.

By the late 1970s, the goals of the institutional-building era had faded as neo-liberal policies were implemented in a growing number of countries and, in the new field of community mental health, cost containment increasingly took precedent over the social integration of former patients as the driver of deinstitutionalisation.[6] Indeed, the widespread practice of transinstitutionalisation, one through which service users—and those who cared for them—were shifted across and between institutional contexts, became characteristic of Canada's evolving welfare state apparatus.[7] As scholars in Canada and elsewhere have already pointed out, the ideological shifts at the high political level created significant challenges, particularly for more vulnerable members of society, including patients as well as underemployed staff associated with these institutions.[8]

One of the goals of the early deinstitutionalisation era was to maintain levels of production and employment in peacetime, as Canada continued to industrialise. Yet even as asylums were downsizing, the emphasis on an individual's value in the workforce continued to frame discourses on categories of ability and disability. Late-era psychiatric institutions attempted to address this trend by replacing work therapy with job and life skills training intended to prepare clients for community living. In a post-asylum world, many people with mental health diagnoses struggled to remain viable members of the labour market. The emergence of sheltered workshops, which Judith Fingard and John Rutherford have investigated in Nova Scotia, fill the role of asylum-based 'work therapy' in a quasi-remunerative environment, sheltered at times from direct labour competition, but subject to economic markets.[9]

[6] See for example, R. Blake's (2009) *From Rights to Needs: A History of Family Allowances in Canada, 1929–1992* (Vancouver: University of British Columbia Press).

[7] C. Dooley (2012) '"The Older Staff, Myself Included, We Were Pretty Institutionalized Ourselves": Authority and Insight in Practitioner Narratives of Psychiatric Deinstitutionalisation in Prairie Canada', *Canadian Bulletin of Medical History*, 29:1, 101–23.

[8] See especially, G. Grob (1994) *The Mad Among Us: A History of the Care of America's Mentally Ill* (New York: The Free Press).

[9] J. Fingard and J. Rutherford (2011) 'Deinstitutionalisation and Vocational Rehabilitation for Mental Health Consumers in Nova Scotia since the 1950s', *Histoire Sociale/Social History*, 44:88, 385–408.

In spite of the structural changes initiated by the closure of long-stay hospitals and the physical impact that destroying large-scale institutions had on the landscapes and economic outlooks for many communities, these dramatic changes did not always bring about significant differences in the experience of alienation or vulnerability faced by many individuals who interacted with the mental healthcare system. Some pharmacotherapies replaced institutional care in an asylum setting but did not fundamentally replace the need for care and attention to disordered, undesirable or unwanted behaviours—whether from the individual's or the physician's perspective.

Deinstitutionalisation may have signalled the end of the age of the asylum and the dawn of new kinds of mental health accommodation, which included a precipitous decline in long-stay patient populations, the gradual closure of separate mental hospitals as psychiatric wards were folded into general hospitals, and the rise of an entirely new system for mental health, but it did not coincide with a decline in the numbers of individuals and families seeking assistance. Quite the opposite occurred. Where the asylum had ostensibly provided a set of services under one roof, problematic though they may have been, the post-asylum world involved a complicated matrix of services. These were not under the jurisdiction of any one governmental department and did not necessarily fit neatly into Canada's constitutional federalist framework. Medical services, housing and employment needs along with financial and family support services required a delicate degree of bureaucratic coordination in a Kafkaesque world of red tape.

For Canadians, the introduction of universal, publicly funded healthcare in the 1960s distinguished them from Americans, who continued (and for now still continue) to add another layer to this bureaucratic entanglement in the form of Health Maintenance Organisations (HMOs) and the healthcare insurance industry. But the introduction of Medicare also significantly affected the way that healthcare services were provided across Canada and, by bringing psychiatric facilities into alignment with the general health system, it produced tensions running from community organisations to legislative assemblies.[10] Successive health commissions have reinforced this perspective, perpetuating an awkward relationship

[10] G. P. Marchildon (2011) 'A House Divided: Deinstitutionalisation, Medicare and the Canadian Mental Health Association in Saskatchewan, 1944–1964', *Histoire Sociale/Social History*, 44:88, 305–30.

between care for the body and care for the mind that continues to challenge service providers and law makers in the mental health field. In his 2002 commission report, *Building on Values: The Future of Health Care in Canada*, Roy Romanow described mental health as the 'orphan child' of Canadian healthcare. He claimed that 'today, mental health care is largely a home and community-based service, but support for it has too frequently fallen short. It is time to take the long overdue step of ensuring that mental health home care services are included as medically necessary services under the *Canada Health Act*, and available across the country.'[11]

Some analyses tend to lament the fall of the asylum, or at least suggest that the new face of mental health has not improved outcomes by plunging service users into a decentralised set of programmes that are underfunded, under-resourced and overpopulated. But other Canadian critics and analysts have stressed the need for user-led or -directed services and embraced, and even demanded, an end to the incarceration of individuals deemed mentally disordered. This sentiment stemmed from a number of cultural movements that had gathered momentum in the midst of rights-based campaigns. Building on the strengths of civil rights, feminism, and gay and lesbian rights movements, radicalised ex-patients in Canada joined fellow activists in the United States, Britain and Europe, and began campaigning for their place in the human rights discourse in the 1970s.[12] Disability rights activists engaged in aggressive campaigns for better access to services,[13] while psychiatric patients and their families began lobbying for anti-stigma campaigns, alongside demands for adequate housing, basic health services, voting rights and access to safe employment. Canadian psychiatric survivor publications, visual art, theatre and businesses flourished in mad-positive urban spaces during the 1980s and beyond, giving birth to a fluid and diverse series of cultural forums.[14] Some of these campaigns and expressions of mad culture were fuelled by or gave inspiration to a set of intellectual critiques that questioned the ways mental disorders

[11] Hon. R. Romanow (2002) *Building on Values: The Future of Healthcare in Canada*, executive summary, p. xxxi.

[12] For a history of the early psychiatric survivor movement in Canada, see I. Shimrat (1997) *Call Me Crazy: Stories from the Mad Movement* (Vancouver: Press Gang).

[13] G. Hudson (2007) 'Regions and Disability Politics in Ontario, 1975–1985', paper presented at 'Region Matters: Health and Place Conference', University of Alberta, Edmonton.

[14] T. Mitchell, D. Dagenais, M. J. Davies, T. McFarlane and I. Shimrat, (2015) 'Survivor Culture', *Après l'Asile/After the Asylum*, http://www.historyofmadness.ca, last accessed 31 Dec. 2015.

were understood and treated. While activists note the empowering aspects of such community involvement, the number of service users who practise cultural and political engagement remains limited. Mental health remains a stigmatised category and Canadian service users rarely seek to celebrate their mental health identities.

A New Perspective on Deinstitutionalisation

When we originally received funding from the Canadian government in 2007 to examine the history of deinstitutionalisation we envisioned a sophisticated, empirically based study of the differences across the country in community care. We were careful to pay attention to gender, race, ethnicity, age and, in particular, rural and urban differences in access to care. Although we believed our work would shake up the Canadian historiography on deinstitutionalisation by emphasising regional differences across the country, we were unprepared for how dramatically our own approach changed over the course of a few years.

Our first major output from the project culminated in a collection of essays in a special issue of a history journal.[15] We congratulated ourselves on pulling together a talented group of scholars who crossed disciplinary lines. Our group included historians, psychiatric nurse historians, a health policy scholar, a psychologist, a sociologist, and one ex-patient. Attempting to position ourselves as sensitive to multidisciplinary perspectives, which we agreed helped to challenge the extant literature in Canada on the topic, we did not yet fully appreciate the significance of bringing an ex-patient to the proverbial table. In fact, as we later reviewed our original research proposal, we realised that we had not even mentioned patients, ex-patients or consumers as targets for research collaboration.

Working with this collection of authors, and digging more deeply into the subject, one article soon took on new meaning for us. Jayne Melville Whyte's examination of the volunteer visitors to the Saskatchewan Mental Hospital at Weyburn was published alongside typical academic pieces, but Whyte's article stood out. She mixed historical excerpts from the Canadian

[15] '*Après l'Asile/After the Asylum*: Réflexions Historiques sur la Désinstitutionalisation au Canada' (2011) special issue in *Histoire Sociale/Social History*, 44:88. This issue was guest edited by Erika Dyck and featured articles from Marie-Claude Thifault, Isabelle Perreault, Geertje Boschma, Kathleen Kendall, Jayne Melville Whyte, Gregory P. Marchildon, Chris Dooley, Catherine Duprey, Judith Fingard and John Rutherford.

Mental Health Association's Manitoba archives with her own candid personal reflections, untethered to scholarly apparatus or a legacy of scholarship. Whyte's analysis of the historical events came from a position of lived experience, though, importantly, not one of memory. Her analytical perspective came from her experiences as a former patient, but her historical subjects stretched back beyond her own experiences, generating a sensitive, intimate and eloquent analysis of this history that unflinchingly took under consideration the sincerity of patients, families and mental health advocates. Whyte encouraged some of us to set aside our academic scaffolding, and instead to stop and listen to different voices both in the archives and in the interpretation of those historical texts.[16]

We tried to respond to her suggestion as we developed research in other regions of Canada. From the beginning of the project, the British Columbia team members had been drawn to the story of Vancouver's Mental Patient Association (MPA), a radical grassroots response to early deinstitutionalisation. Formed in early 1971 by former psychiatric patients and their allies, MPA was the first service-user group to emerge in Canada.[17] Its founders created a politically active and uniquely successful support system that provided work, homes and a sense of belonging and empowerment to ex-patients. Established in an era of grassroots community initiatives and political radicalism, the early MPA dramatically changed the way community participation was conceptualised; it actually modelled how to create community. It remains a powerful model of best mental health practice and an inspiration to those who believe in user-directed services.

Discovering that the current MPA had retained no historical records beyond copies of the colourful *In A Nutshell* tabloid produced by the

[16] Whyte's work is not the only work that has inspired this kind of historical scholarship, yet it was critical for our own journey. See, for example, G. Reaume (2000) *Remembrances of Patients' Past: Patient Life at the Toronto Hospital for the Insane, 1870–1940* (Toronto: Oxford University Press).

[17] Literature on the history of the MPA includes: G. Boschma, M. J. Davies and M. Morrow (2014) '"Those People Known as Mental Patients ...": Professional and Patient Engagement in Community Mental Health in Vancouver, BC in the 1970s', *Oral History Forum*, 34; M. J. Davies, G. Boschma and M. Morrow (2015) 'How MPA Re-formed Community Mental Health', *Après l'Asile/After the Asylum*, http://www.historyofmadness.ca, last accessed 31 Dec. 2015; J. Chamberlain (1979) *On Our Own: Patient Controlled Alternatives to the Mental Health System* (New York: McGraw-Hill); I. Shimrat (2013) *Call Me Crazy*; *The Inmates Are Running the Asylum: Stories From MPA*, DVD and online, directed by the MPA Documentary Collective (Canada: History of Madness Production).

group, team members Geertje Boschma, Marina Morrow and Megan J. Davies conducted a series of videotaped interviews with early MPA members, community allies and local psychiatrists whose professional lives had intersected with the establishment of community mental health in the western city. The rich stories and important insights collected inspired the academics to abandon a standard scholarly research approach and invite a group of eleven early MPA members to join in the creation of a website exhibit on the organisation.[18] When the group gathered for the first time, however, the choice of the MPA interviewees was to use their oral history footage to create a documentary about the vibrant story of MPA's first decade.[19] The decision to create a publicly accessible research product came to characterise the After the Asylum projects and a different process quickly became apparent in the creation of *The Inmates Are Running the Asylum: Stories from MPA*.[20] Interviewees reviewed text copies of their videotaped oral histories, highlighting sections *they* considered 'important', a flexible term that left historical judgement in the hands of those who had participated in it. Gathering the interview selections and using her historical training, Davies created topical and thematic clusters, placed them on a rudimentary timeline/documentary storyboard, and took it back to the group for review. While Davies, Morrow, and Boschma assumed the shared role of producer, the MPA Founders directed the documentary.

The Inmates emerged as a collaborative project that shared ownership of history, and in the process addressed questions of voice, interpretation, responsibility and audience.[21] The participants brought their expertise to

[18] The MPA documentary group included seven people who identified as former psychiatric patients, and two allies.

[19] The MPA founders who worked on the film include: Ian Anderson, Dave Beamish, Lanny Beckman, Avi Dolgin, Patty Gazzola, Arthur Giovinazzo, John Hatfull, Jackie Hooper and Alex Verkade.

[20] *The Inmates* can be viewed at: https://www.youtube.com/watch?v=JwyaRU1svrA; MPA Documentary Collective (2015) 'The Inmates Are Running the Asylum', *Après l'Asile/ After the Asylum*, http://www.historyofmadness.ca, last accessed 31 Dec. 2015; HiP/HeT project team (2016) 'MPA: The Inmates' Utopia/ MPA: L' Utopie des Internés', *History in Practice: Community-Informed Mental Health Curriculum/Histoire en Tête: Sagesse Communautaire et Apprentissages*, http://historypractice-histoireentete.ca/, last accessed 30 Apr. 2016. The HiP/HeT site includes bilingual French/English teaching resources and a version of the film with French subtitles. DVD copies of *The Inmates* are available from the producer upon request.

[21] Publications which address these issues include: K. Borland (1991) '"That's Not What I said": Interpretive Conflict in Oral Narrative Research' in S. B. Gluck and D. Patai (eds)

the table, as people with living memories, as organisational facilitators, as artists, storytellers, analysts and editors. Working together, academic and community perspectives dissolved into a fluid mixture of ideas and critiques governed by the principle of participatory democracy—a founding MPA tenet.[22]

Perhaps because of their interdisciplinary backgrounds, their tenured status and the nature of their work in healthcare, the three academics on the project were comfortable letting go of feelings of academic ownership over the outcomes. The MPA Founders determined how history was presented in the film, a hallmark of the field of oral history and an extension of the notion of shared authority.[23] 'The real responsibility', Davies wrote at the time, 'is not just to get the story right in academic terms, but to facilitate both a process and a product that [survivors] believe to be authentic'.[24] The experts by experience provided an interpretation of the past that was informed both by a deep longing and a sense of present purpose: as community expert John Hatfull stated at one meeting, 'I have the

Women's Words: The Feminist Practice of Oral History (New York: Routledge) pp. 63–71; R. Kearns (1997) 'Narrative and Metaphor in Health Geographies', *Progress in Human Geography*, 21:2, 269–77; J. Sangster (1994) 'Telling Our Stories: Feminist Debates and the Use of Oral History', *Women's History Review*, 3:1, 5–28; A. Sheftel and S. Zembrazycki (eds) (2013) *Oral History Off the Record: Towards an Ethnography of Practice* (New York: Palgrave Macmillan).

[22] Francesa Polletta took this approach in her oral history work with veteran community activists in the United States, modelling her research methods on the organisational methods that had been employed by her subjects in the 1960s. See F. Polletta (2002) *Freedom is an Endless Meeting: Democracy in American Social Movements* (Chicago, IL: University of Chicago Press). For a fuller explanation of the process of making *The Inmates* see L. Beckman and M. J. Davies (2013) 'Democracy is a Very Radical Idea', in B. A. LeFrançois, R. Menzies and G. Reaume (eds) *Mad Matters: A Critical Reader in Canadian Mad Studies* (Toronto: Canadian Scholars' Press), pp. 49–63.

[23] Boschma trained as a nurse and a historian. Morrow trained as a community psychologist but works primarily on health policy. Davies trained as a historian but has taught in health studies for more than a decade. Presented first in 1990 by American historian Michael Frisch as operating within the confines of the interview process, the concept of shared authority has evolved and expanded in meaning to include the entire oral history process which is understood to include self-reflection on the part of the researcher and the affirmation of self and community of among those who are interviewed. M. Frisch (1990) *A Shared Authority: Essays on the Craft and Meaning of Oral and Public History* (Albany: State University of New York Press); M. Frisch (2003) 'Sharing Authority: Oral History and the Collaborative Process', *Oral History Review*, 30:1, 111–13; S. High (2009) 'Shared Authority: An Introduction', *Journal of Canadian Studies*, 43:1, 12–34.

[24] Beckman and Davies, 'Democracy is a Very Radical Idea', *Mad Matters*, p. 60.

vain hope that if we talk about it enough maybe we will find a way to recreate the MPA the way it was. I always thought it was a great idea, I loved it, and I thought it helped more people than any other group.'[25] Hatfull's motivation for participation helped to illustrate the contemporary relevance of the project and reinforced how deinstitutionalisation continues to function as a living history, co-interpreted by those who study the history and by those who experienced it directly.

Co-created Knowledge

As *The Inmates* documentary project makes clear, generating co-created deinstitutionalisation histories involves borrowing approaches and methodologies from oral history and survivor research, and begins to incorporate subjects directly into the decision-making process regarding interpretation, analysis and dissemination.[26] One of the key ingredients of the kind of activist history we have been making in the various After the Asylum projects on the website is a redefined understanding of expertise whereby those whose lives have been shaped by deinstitutionalisation and its legacies are understood to be more than just people who hold relevant stories, becoming integral project players in deciding what to emphasise, who to tell, and how to present this complex and sometimes troubling history.

To explore the way in which the After the Asylum projects have reworked the notion of expertise, we developed the History in Practice: Community-Informed Mental Health Curriculum/Histoire en tête: sagesse communautaire et apprentissages (HiP/HeT) educational website, which drew on the After the Asylum research to create an innovative set of teaching tools for training future mental health professionals. Bringing the history of deinstitutionalisation to new audiences, from the outset HiP/HeT was intended to be a collegial and continuing collaboration between

[25] M. J. Davies, Project notes, MPA Documentary Meeting, 6 April 2011.

[26] Sharing authority involves opening up possibilities for a collective conversation and a project that allows all parties to expand their skill set and knowledge base: oral historians now involve storytellers in determining project design, protocols and products. See D. Kerr (2003) '"We Know What the Problem Is": Using Oral History to Develop a Collaborative Analysis of Homelessness from the Bottom Up', *The Oral History Review*, 30:1, 27–45. In survivor research there is a keen appreciation for the power dynamics of owning authoritative knowledge. See A. Sweeney, P. Beresford, A. Faulkner, M. Nettle and D. Rose (eds) (2009) *This is Survivor Research* (Ross-on-Wye, UK: PCCS Books).

academics and people with direct experience of receiving mental health services and support.

Broadening the focus in After the Asylum to include contemporary mental health practice brought an appreciation by non-historians and community partners alike for 'real world' applications of project research findings, uses which directly influence policy and practice while improving peoples' lives. Education therefore quickly became identified as a key target for dissemination of project findings.[27] The belief of HiP/HeT service users in the potential of the project was particularly evident. 'We're just hoping there'll be a wave of change', wrote community experts Dana Allen and Alistair Scott-Turner, 'and the doctors and psychiatrists, maybe they'll start treating people the way they would want to be treated'.[28] Alex Verkade joined the project because he was interested in demonstrating, 'that mental health consumer/survivors can contribute in bettering their own situation'.[29]

Project community experts used brainstorming sessions and small group discussions to identify key educational themes. Team historians mined the academic data, selecting work that had strong historical merit but would also provide engaging, teachable intersections between past and present.[30] For instance, Deborah Thien, a geographer with research expertise in the study of emotion and mental health, proposed that the Respecting Emotions Module would comprise an interactive web map of Canadian emotions in the deinstitutionalisation era. In this fashion the academics on the project did not surrender their position as experts, but acknowledged that there was another kind of expert knowledge at play

[27] Two secondary school History of Madness/Histoire de la Folie educational sites were also created using *Après l'Asile/After the Asylum* research: M. J. Davies and E. A. Marshall (2009) *Caring Minds: Youth, Mental Health and Community*, http://www.caringminds.ca, last accessed 2009 and M. J. Davies and D. Purvey (2010) *More for the Mind: Histories of Mental Health for the Classroom*, http://www.historyofmadness.ca, last accessed 2010.

[28] D. Allen and A. Scott-Turner (2016) 'The Psychiatric Gaze', *History in Practice: Community-Informed Mental Health Curriculum/Histoire en Tête: Sagesse Communautaire et Apprentissages*, http://historypractice-histoireentete.ca/, last accessed 30 Apr. 2016.

[29] A. Verkade and C. Wall (2016) 'Unity Housing: Model of Best Practice/ Space of Hope', *History in Practice: Community-Informed Mental Health Curriculum/Histoire en Tête: Sagesse Communautaire et Apprentissages*, http://historypractice-histoireentete.ca/, last accessed 30 Apr. 2016.

[30] The six teaching units on the site include: Medicine and Power, Models of Practice (Spaces of Hope), Policy into Practice, Stigma and Discrimination, Advocacy and Activism, and Respecting Emotions.

and looked for ways to incorporate multiple interpretations and frameworks for learning into the teaching resources. Buoyed by the success of this first step, the group agreed that the community experts would create their own written and visual materials to further illuminate the history of post-asylum mental health. The result was a rich array of historical documents and images, oral histories, video, art, creative writing and memoirs; a diverse set of learning 'artefacts' which, as project member Marie-Claude Thifault notes, will help students move beyond the taboos associated with mental health and open their minds to new perspectives.[31] The site went live in winter 2016/2017 (http://historypractice-histoireentete.ca), and a national email campaign targeted to potential programme designers and educators leads recipients to a promotional video on the site home page.

Two units from the Medicine and Power Module demonstrate the value of HiP/HeT's process of co-created knowledge. 'The Doreen Befus Story: The Law and Practice of Sterilisation', uses a biographical approach and research conducted by historian Erika Dyck to tell the story of Alberta's provincial sterilisation programme, in operation from 1928 to 1972. [32] In a second Medicine and Power unit, community expert Irit Shimrat developed a 1978 psych ward notebook of sketches and prose into 'Question Man', a six-minute video which tells how this alter-ego and friend helped Shimrat reclaim a creative spirit and sense of self. In the culminating learning activity to the 'The Doreen Befus Story', students use a Sterilisation Timeline which sets details from Doreen's life alongside a national and international history of eugenics, sterilisation and human rights, adding their own annotations drawn from unit documents, images and audio commentary to illuminate the thoughts and responses of Befus, other victims/activists, lawmakers and medical staff. After viewing 'Question Man', learners are challenged to consider how to work with service users to assist people with mental health difficulties to locate knowledge and strategies within themselves.

Powerful artefacts and learning support materials like these reflect project community experts' wishes that future mental health practitioners receive a strong message about the professional authoritarianism, lack of

[31] Project promotional video (2016) *History in Practice: Community-Informed Mental Health Curriculum/Histoire en Tête: Sagesse Communautaire et Apprentissages*, http://historypractice-histoireentete.ca/, last accessed 30 Apr. 2016.

[32] Dyck wrote an account of Befus' life story for the *Après l'Asile/After the Asylum*. For a broader academic account, see E. Dyck (2013) *Facing Eugenics: Sterilisation, Reproduction and the Politics of Choice* (Toronto: University of Toronto Press), ch. 6.

holistic forms of treatment, and the need for respectful communication that they believe is characteristic of the current mental health system. By having future mental health practitioners evaluate the laws and practices of sterilisation in Alberta and one victim's response, and consider how they might work in a healing partnership with a service user, the craft of history exposes students to a set of analytical tools for navigating the complex world of the twenty-first century mental health system. As Canadian historian and educator Ruth Sandwell states, 'historical thinking in all of its complexity provides ... the broad humanistic knowledge of, and the best ways to create and critique truth-based claims within, a pluralist participatory democracy'.[33] Like other HiP/HeT teaching materials, these units foster an informed empathy, encouraging learners to consider people, events and situations in their context.[34] In this kind of activist history, knowledge is co-created by academic researchers, community experts and students. HiP/HeT functions as a history of deinstitutionalisation on the move. It provides multiple perspectives and access points as well as a variety of dissemination goals that reach a wider swath of Canadians than traditional academic products.

The Power of the Story

'Activism, like history, begins with the telling', radical historian Max Page wrote in 2001, arguing that 'memory-work', and giving a voice to remembrance, is a fundamental democratic act.[35] Activist public histories, like the multiple projects to which After the Asylum gave birth, endeavour to share ownership of the past, and by doing so 'trouble' standard

[33] R. W. Sandwell (2011) 'On Historians and their Audiences: An Argument for Teaching (and not just Writing) History', Address to the 'Where the Archive Ends: A Graduate Conference on History and Its Uses', McGill-Queen's University Graduate Conference, Kingston, 11 Mar. 2011. S. Levesque (2008) *Thinking Historically: Educating Students for the Twenty-First Century* (Toronto: University of Toronto Press), provides a particularly good summary of the literature relating to this point in his introduction. See also P. Seixas (Dec. 2009) 'A Modest Proposal for Change in Canadian History', *Teaching History*, 137, 26–30.

[34] For an exploration of empathy in the practice of history and education see O. L. Davis (2001) 'In Pursuit of Historical Empathy' in O. L. Davis, E. A. Yeager and S. Foster (eds) *Historical Empathy and Perspective Taking in the Social Studies* (Lanham, MD: Rowland and Littlefield), pp. 1–12.

[35] M. Page (Winter 2001) 'Radical Public History in the City', *Radical History Review*, 79, 114–16.

scholarly ideas about voice, interpretation, responsibility and audience.[36] This is humanistic history practice, which allows new sets of relationships and research agendas to emerge and, as historian Ludmilla Jordanova has observed, sheds light on remnants of the past that are everywhere but are not regarded as history.[37] This is certainly true in the mental health world. Most members of the public—and many mental health practitioners and patients—lack a nuanced understanding of the evolution from residential to community care. Project health historian Chris Dooley's online exhibit, 'Working Between Mental Health Worlds: Practitioner Voices on Deinstitutionalization', contains a sensitive presentation of the voices of practitioners who worked in old line mental health institutions, then followed their former patients into community settings.[38] The stories of the former practitioners demonstrate that while the institutions themselves largely disappeared, institutional treatment models, characterised by paternalism and control, persisted, following their subjects into the new community settings.

Using a set of oral histories collected from registered psychiatric nurses in Saskatchewan and Manitoba, Dooley locates psychiatric deinstitutionalisation within a wider project of transinstitutionalisation.[39] Focusing on these liminal spaces and liminal bodies that moved among institutions, for work and for care, Chris Dooley explores the evolution of mental health provision at the height of 1960s and 1970s deinstitutionalisation in two prairie provinces, tracking practitioners whose careers began in the 1950s and 1960s at large inpatient facilities and who later worked in community roles. He presents not so much a well ordered set of memories, but rather a series of recollections and personal reflections, some of which challenge commonly held stereotypes and assumptions about mental health workers.

Deinstitutionalisation had profound effects on the lives of frontline mental health workers. As the big institutions, such as the Weyburn,

[36] For example, see S. High (2014) *Oral History at the Crossroads: Sharing Life Stories of Survival and Displacement* (Vancouver: University of British Columbia Press); and P. Sugiman (2009) '"Life is Sweet": Vulnerability and Composure in the Wartime Narratives of Japanese Canadians', *Journal of Canadian Studies*, 43:1, 186–218.

[37] L. Jordanova (2006) *History in Practice* (London: Hodder Arnold). We are working here with ideas about humanistic oral history practice discussed in A. Sheftel and S. Zembrazycki (eds) (2013) *Oral History Off the Record: Towards an Ethnography of Practice* (New York: Palgrave Macmillan), pp. 7–8.

[38] http://aftertheasylum.apps01.yorku.ca/en/workingBetweenMentalHealthWorlds.

[39] Dooley, 'The Older Staff, Myself Included, We Were Pretty Institutionalized Ourselves'.

North Battleford, or Brandon mental hospitals, closed in the 1960s and 1970s, many practitioners shifted to working as community mental health nurses with social services agencies, or on the acute care psychiatric wards that opened in general hospitals. These were fraught times when mental health workers faced professional and economic dislocation and uncertainty. As jobs disappeared in these old 'asylum towns', some were even forced to uproot their families and move elsewhere. New contexts offered new challenges and practitioners were obliged to develop different skills and approaches, to reinvent themselves as mental health care providers, and to re-examine patient–practitioner relationships.

Taking account of these dislocations helps us to see practitioners not simply as instruments of a bureaucratised health and welfare system, but as individuals implicated in complex relationships with the people and systems they served. By paying attention to their own words, we can appreciate that many imagined themselves not simply as professional caregivers, but also as protectors, allies, advocates and friends to those for whom they cared. Although deinstitutionalisation presented personal and professional challenges, most of the practitioners interviewed embraced the goals of care in the community. One woman summed it up in seven words: 'It was necessary; it was their [patients'] right.' All, however, expressed reservations about the means by which this was achieved, and this same woman, after a long, emotional pause, followed with the qualification, 'But sometimes we went too far, too fast.'[40] Another respondent said, 'Closing the mental hospitals was always my ultimate goal, but I thought that it was too soon .... I think we dumped, and I don't believe in dumping.'[41] These realisations confirm established arguments about the pace of deinstitutionalisation, which was particularly accelerated in Saskatchewan, as well as the often ambivalent response of practitioners to the impersonal nature of this seismic shift in public policy.

Most often, these reservations came from the sense that deinstitutionalisation was driven by political expediency, and practitioners condemned what they saw to be the inhumane policies. Hospitals were downsized during the same period as residential schools for Aboriginal children closed their doors, signalling the end of an era of large-scale rehabilitation institutions, but also

[40] Jane S. Interview by Chris Dooley, Battleford SK, 2 Apr. 2006.
[41] Pat G. Interview by Chris Dooley, Brandon MB, 21 Apr. 2007.

the dawn of a new era of fiscal conservatism.[42] 'When the big hospital emptied, the patients were thrown out and there was no machinery in place. Nothing was done for them. [They] just threw them out on the street, filled them with tranquillizers. And that generation just died', one former asylum worker told Dooley.[43] As practitioners moved away from traditional custodial institutions and into community mental health positions or into non-nursing roles with community agencies, they gained insight into some of the problems of the older system. Some recalled that it was only once they were established in the community that they appreciated the nature of confinement. Some commented on how they now saw ex-patients as people. One person recalled, 'You're quite blinded when you only work with inpatients. You only see a patient. Only when I became a CPN [Community Psychiatric Nurse] did I see a person.'[44]

Others indicated that working in the community gave them the opportunity to re-evaluate their role in prior regimes of paternalism and authoritarian control. One nurse reflected that 'out in the community, the power relations were totally different. OK, they weren't really reversed, because you still had the power, but when you went to see them, it was like you were a guest in their home. And that made a big difference.'[45] Another recalled that

> We used to joke and brag that the bars were there to keep the discharged patients from climbing back in the windows, but even those who struggled and had tough times seldom wanted to come back. In the end I took this as a sign that whatever kind of [institutional] community we provided, for most it was never really home to them.[46]

Moving to a new vantage point in the community altered the nurses' interpretations of the relationships within the system between themselves and the patients and ex-patients.

Unsurprisingly, Dooley found many of these practitioners remained active in mental health education and advocacy well into their retirement. In these roles, though, they sometimes felt alienated by shifts in mental health

[42] For more on residential schools, see J. Miller (1996) *Shingwauk's Vision: A History of Native Residential Schools* (Toronto: University of Toronto Press).

[43] Len S. Interview by Chris Dooley, Debden SK, Brandon MB, 5 Apr. 2006.

[44] Karin K. Interview by Chris Dooley, Saskatoon SK, 6 Apr. 2006.

[45] Nevin S. Interview by Chris Dooley, Yorkton SK, 8 Apr. 2006.

[46] Ben K. Interview by Chris Dooley, North Battleford SK, 30 Mar. 2006.

activism that saw practitioners and former practitioners marginalised within activist circles.[47] The rise of mental health consumerism, with its emphasis on survivor self-determination, has been a source of ambivalence for many of these practitioners who, in their working lives, often self-identified as being in the progressive vanguard of mental health activism and who are now sometimes labelled as apologists of the old, paternalistic system.

Often, the practitioners' stories about deinstitutionalisation became a professional reflection on working between two mental health worlds. They described how they came to recognise the paternalism of the asylum system and their own role in it, but also their own misgivings about the process of deinstitutionalisation: 'On reflection, many came to see that their own expectations of patient behaviour and potential were framed by practices that cultivated and rewarded chronicity and dependency and discouraged self-determination.'[48] As one oral history respondent observed:

> The hardest transition for me was from nurturing and fixing to stepping-back and letting you solve your own problems. We thought that we were helping [people] by not letting them make bad decisions; in the end, the bad decision was not letting go. Letting go is the hardest part; you've got to let people succeed or fail on their own terms.[49]

Such reflections underscore the importance of tapping the retrospective insights of current and former practitioners to better understand the complexities of deinstitutionalisation. It makes for a messy history—one freighted with conflated timelines and the confusion of contemporary and contemporaneous opinion—but in its commitment to sharing interpretative authority, it opens the possibility for a vital democratisation of the process of making history.

Conclusion

In writing about the history of deinstitutionalisation we now recognise that the Canadian story is enriched by embracing a community-engaged ethos, by paying attention to the myriad people, institutions and ideas that have been affected by the decision to shift the weight of the mental

[47] For a discussion of this alienation, see W. R. Curtis (1986) 'The Deinstitutionalization Story', *Public Interest*, 85:1, 34–49, esp. 44.

[48] Dorie T. Interview by Chris Dooley, Prince Albert SK, 6 Apr. 2006.

[49] Ibid.

health system from an institution-based model, to one ostensibly governed by principles of self-determination, though often characterised by the realities of an impoverished system. By examining the history of deinstitutionalisation from the perspective of those with lived experience, both staff and patients, and by generating historical products aimed at reaching beyond academic audiences we contribute to a bigger project of public memory and mental health, with deinstitutionalisation as the entry point for discussion.

The After the Asylum website and its associated projects are about rethinking perspective, process and product; and we encourage historians to reflect on the process of working collaboratively with people who often exist at the margins of our society, and who have more often been the objects of study than the authors of their own past. These various initiatives ultimately involved approximately 90 team members—academic researchers, community partners, and a host of research assistants, artists and technicians. Community partnerships have turned out to be a significant part of these projects. The Mental Patients Association Founders want their film to be presented widely as a 'citizen's resource', a piece of active history which invites the viewer to revaluate her or his viewpoint about mental health and community formation.[50] The film makes public a historical moment when hierarchies were reordered, when the inmates were running the asylum and when the academics held the cameras. Making the film created a new authorial discourse on this topic, where only scattered remnants existed before in a thin documentary trail.

The voices, contributions and perspectives of people connected with this history as service users lies at the heart of our studies. Within the first few years of developing our respective projects, persons with lived experience of mental health difficulties became more than just project research subjects, emerging as collaborators and co-authors. We began practising a different kind of history, initiating a process where academic team members took on the roles of listeners, learners and facilitators, while elevating the roles of community actors and letting them set much of the research agenda. Working alongside our community partners, we began to co-create a history of deinstitutionalisation closely connected to community and with a strong public purpose.

[50] The phrase 'citizen's resource' comes from Tosh, *The Pursuit of History*, quoted in J. Kalela (2012) *Making History: The Historian and the Uses of the Past* (Basingstoke: Palgrave Macmillan), p. 61.

As we developed these approaches to the research and authorship, our end products also changed. Projects became diverse, growing from a thematic academic journal issue to include a documentary film, an interactive website with exhibits that range in terms of academic language or artistic expressions of madness, and an innovative teaching website that brings together the cultural products of this work and employs history to train future mental health practitioners. At its core, our individual and collective exhibits work at disrupting tidy narratives, academic discourse and top-down analyses, which we believe reinforce disempowering traditions that treat mental patients as incapable of full participation in society, including contributing to how we remember the history of psychiatric institutionalisation.

PART II

Designing and Implementing Psychiatric Reform: Experiments, Opportunities and Oppositions

CHAPTER 5

'Islands of Reform': Early Transformation of the Mental Health Service in Lower Saxony, Germany in the 1960s

Christof Beyer

After the Second World War, the mental health care system in Germany was in a desolate state. Approximately 300,000 psychiatric patients lost their lives between 1939 and 1945: they were killed in gas chambers as part of the so-called 'euthanasia' programme of the National Socialists; they died from medication overdoses, intentional neglect and hunger; and they were executed by special command of the SS (Schutzstaffel) in Poland and mobile killing units (Einsatzkommandos) in the Soviet Union.[1] Furthermore, at least 300,000 people had become victims of compulsory sterilisation by the so-called 'Law for the Prevention of Genetically Diseased Offspring' (Gesetz zur Verhütung erbkranken Nachwuchses) enacted in 1933.[2]

[1] H. Faulstich (2000) 'Die Zahl der "Euthanasie"-Opfer' in A. Frewer and C. Eickhoff (eds) *'Euthanasie' und die Aktuelle Sterbehilfe-Debatte. Die Historischen Hintergründe Medizinischer Ethik* (Frankfurt a.M./New York: Campus-Verlag), pp. 218–34.

[2] H. Friedländer (1995) *The Origins of Nazi Genocide. From Euthanasia to the Final Solution* (Chapel Hill: University of North Carolina Press), pp. 25–30.

C. Beyer (✉)
Berlin, Germany
e-mail: christofbeyer@gmx.de

D. Kritsotaki et al. (eds.), *Deinstitutionalisation and After*,
Mental Health in Historical Perspective,
DOI 10.1007/978-3-319-45360-6_5

Psychiatric hospitals were mostly in a disastrous condition due to radical budget cuts that dated back to the 1930s, their conversion into other functions such as military hospitals during the 1940s, and bombings.[3] Most of the psychiatrists, including clinical directors and heads of university departments, had been involved in National Socialist actions against people with mental illnesses and disabilities. Under the 'Third Reich', Jewish psychiatrists had been banned from their profession, with many fleeing to North America and elsewhere. Racial hygiene and eugenics dominated research activities as they were aimed at 'verifying the world view of the national socialists'.[4] Psychiatrists at university departments as well as doctors in psychiatric hospitals used 'euthanasia' killings of children for their own research or even expedited murders—one example of the latter was Professor Carl Schneider at the University of Heidelberg who examined the brains of children killed in the Eichberg psychiatric hospital.[5] After the war, only a few of these doctors were brought to justice and convicted of their crimes.[6] Some of the psychiatrist perpetrators, among them Carl Schneider, committed suicide.[7]

Such is the background for the development of the mental health care system in Germany after 1945. In the early years of the Federal Republic of Germany founded in 1949, most of the energy of professionals went into rebuilding the asylums. As the German historian Hans-Walter Schmuhl wrote, the psychiatric practitioners 'drew themselves back behind the walls of run-down asylums, which soon became overcrowded again'. One reason for this high occupancy was that a lot of German asylums had been closed or destroyed during the Second World War; another reason was the lack of therapeutic alternatives to

[3] C. Brink (2010) *Grenzen der Anstalt. Psychiatrie und Gesellschaft in Deutschland 1860–1980* (Göttingen: Wallstein-Verlag), pp. 360–65.

[4] W. U. Eckart (2012) *Medizin in der NS-Diktatur. Ideologie, Praxis, Folgen* (Wien: Böhlau), p. 281.

[5] M. Rotzoll and G. Hohendorf (2012) 'Krankenmord im Dienst des Fortschritts? Der Heidelberger Psychiater Carl Schneider als Gehirnforscher und "Therapeutischer Idealist"', *Nervenarzt*, 3 (2012), 311–20, pp. 317–20.

[6] H. Loewy and B. Winter (eds) (1996) *NS-'Euthanasie' vor Gericht. Fritz Bauer und die Grenzen Juristischer Bewältigung* (Frankfurt a.M./New York: Campus-Verlag).

[7] I. Harms (2010) 'Die Gutachter der Meldebogen. Kurzbiografien', in M. Rotzoll, G. Hohendorf, P. Fuchs, P. Richter, C. Mundt and W. U. Eckart (eds) *Die Nationalsozialistische 'Euthanasie'-Aktion 'T4' und ihre Opfer* (Paderborn: Schönigh), pp. 405–20.

the psychiatric hospitals.[8] In post-war Germany of the 1950s and 1960s, psychiatric hospitals were the 'poorhouses of medicine';[9] psychiatric patients ranked low in the welfare system and were, regarding the financial support of health insurances and the possibilities of rehabilitation, disadvantaged in comparison to physically handicapped and physically ill people. In fact, there were immense supply gaps with no professional support between the big psychiatric hospitals, mostly situated in rural areas far away from the cities, and very few psychiatric specialists with their own independent practices. In combination with the 'chemical revolution' of the newly available neuroleptic drugs in the 1950s, this led on the one hand to the 'revolving door syndrome' of patients with lots of short-term readmissions to hospitals.[10] On the other hand, large clinics also had to deal with many so-called 'long term patients', who had been hospitalised often for years or even decades.

In this situation, a few young psychiatrists began to try to establish new, innovative forms of mental health care. As they had started and completed their degrees in medicine after the Second World War, they had not been involved in psychiatric crimes under the Third Reich, as was the case with many others still working as practitioners and university professors in Germany in the 1950s and 1960s. These young psychiatrists were highly aware of the critical situation of German psychiatry and its inhumane past. By the beginning of the 1960s, the first 'islands of reform'[11] started to develop, and the department of psychiatry at Heidelberg University where Karl Peter Kisker started his career was one of them.

[8] Brink, *Grenzen*, p. 376; H. Faulstich (2003) 'Die Anstaltspsychiatrie unter den Bedingungen der "Zusammenbruchsgesellschaft"' in F. W. Kersting (ed.), *Psychiatriereform als Gesellschaftsreform. Die Hypothek des Nationalsozialismus und der Aufbruch der Sechziger Jahre* (Paderborn: Schöningh), pp. 21–30, p. 22.

[9] H. W. Schmuhl (1993) 'Kontinuität oder Diskontinuität? Zum epochalen Charakter der Psychiatrie im Nationalsozialismus' in F. W. Kersting, K. Teppe and B. Walter (eds) *Nach Hadamar. Zum Verhältnis von Psychiatrie und Gesellschaft im 20. Jahrhundert* (Paderborn: Schöningh), pp. 112–136, p. 135.

[10] W. Rudloff (2008) 'Rehabilitation und Hilfen für Behinderte' in M. Ruck and M. Boldorf (eds) *Geschichte der Sozialpolitik in Deutschland seit 1945, Bd. 4. 1957–1966. Bundesrepublik Deutschland: Sozialpolitik im Zeichen des Erreichten Wohlstandes* (Baden-Baden: Nomos Verlag), pp. 463–501, p. 494.

[11] Rudloff, 'Rehabilitation', p. 501.

Karl Peter Kisker, born in 1926, had joined the military service in 1943 as a 17-year-old boy. After the war, he first studied theology and then shifted to medicine and psychology at Heidelberg University. He received a Doctor of Medicine with a thesis on epileptic fugues in 1952 and also a Doctor of Philosophy with a thesis on short-sightedness in 1955. In the same year, Kurt Schneider, the chief physician and head of the psychiatric-neurologic clinic at the university, was succeeded by Walter Ritter von Baeyer. Kurt Schneider had become famous for his 1950 work *Klinische Psychopathologie* (*Clinical Psychopathology*), which established the definition of first-rank and second-rank symptoms for schizophrenia and had an important impact on the third edition of the influential *Diagnostic and Statistical Manual of Mental Disorders* (1980), published by the American Psychiatric Association. Schneider's system of clearly defined classifications of psychiatric diseases was based on objectively verifiable symptoms, radically separating the form and content of psychotic phenomena. His successor in Heidelberg, Walter Ritter von Baeyer, later criticised this model for its lack of comprehension of psychoanalytical approaches to the unconscious and of overall anthropological aspects, regarding it as 'conceptually isolating', 'distant from body and society' and 'deliberately objectifying'.[12] In 1949, Baeyer visited the United States and was 'overwhelmed' by the 'confusing diversity' of the psychiatric system.[13] He described the condition of many US state psychiatric hospitals as below the 'average level' in Europe but was at the same time impressed by approaches of depth psychology to psychoses implemented by—in his opinion—mainly German and Austrian psychoanalysts. Baeyer found them just as fruitful as the move towards a 'therapeutic community' he learned about at the Menninger Clinic and the 'non-directive therapy' (later named 'person-centred therapy') formulated by Carl Rodgers.[14] With Baeyer's appointment as the new head in 1955, the department of psychiatry at Heidelberg University shifted from a psychopathological school towards anthropological approaches to psychiatry and an emphasis on social aspects of psychiatry.[15]

[12] W. v. Baeyer (1977) 'Walter Ritter von Baeyer' in L. J. Pongratz (ed.) *Psychiatrie in Selbstdarstellungen* (Bern: Huber), pp. 9–34, p. 12.

[13] On the development of the psychiatric system in the United States see for example M. Smith (2008) 'Psychiatry Limited: Hyperactivity and the Evolution of American Psychiatry, 1957–1980', *Social History of Medicine*, 3, 541–59.

[14] Baeyer, 'Baeyer', p. 23; W. v. Baeyer (1950) 'Gegenwärtige Psychiatrie in den Vereinigten Staaten', *Nervenarzt*, 1, 2–9.

[15] M. Rotzoll (2011) *Das Zentralinstitut für Seelische Gesundheit. Die Gründungsgeschichte—Aufbruch in die Psychiatriereform* (unpublished manuscript, Heidelberg University), p. 38.

In 1959, Karl Peter Kisker qualified as university lecturer in neurology and psychiatry with the topic *Der Erlebniswandel des Schizophrenen* ('The Change in Experience of the Schizophrenic').[16] In this work he focused on an anthropological understanding of schizophrenia and defined psychiatry as a philosophical discipline of medicine. Thus, he was in opposition to the German mainstream of psychiatric practice in the 1950s, which mainly focused on traditional somatic approaches to mental illness as represented by Kurt Schneider's 'clinical psychopathology', Karl Kleist's 'brain pathology' and Ernst Kretschmer's 'constitutional psychiatry'.[17] In contrast to this approach of individual pathology, Kisker regarded schizophrenia as a pathological condition of human relationships, defining the schizophrenic patient as fallen out of dialogue. Therefore, the dialogue with the 'madmen' had to be resumed to train the co-existence with mentally deviant people through patient and persistent reflection.[18] In this way, Kisker's work followed new 'anthropological', 'phenomenological' and 'existential analytical' approaches of psychiatry, which 'marked a first phase of criticism' on traditional psychiatry in Western Europe and the United States in the 1950s.[19] Kisker regarded mental illness as a 'positive possibility of being human which affects everyone'.[20] Thus, his concept of schizophrenia was quite similar to R. D. Laing's model elaborated in his book *The Divided Self*, which was published a year before Kisker's 'Erlebniswandel'. Laing's approach of 'existential phenomenology' was—like Kisker's—based on the work of Martin Heidegger and Ludwig Binswanger's 'existential analysis'.[21] Therefore, Kisker's psychiatric approach was hermeneutic and accepting, and it formed the basis for his future practical work in Heidelberg and Hannover.[22] As far as we know, there is no evidence about

[16] W. Machleidt (2007) 'Zur Biografie Karl Peter Kiskers' in W. Machleidt, T. Passie and D. Spazier (eds) *PsychiaterSein. Karl Peter Kisker—Auswahl seiner Schriften* (Bonn: Psychiatrie-Verlag), pp. 364–66.

[17] H. W. Schmuhl (1993) 'Kontinuität oder Diskontinuität? Zum Epochalen Charakter der Psychiatrie im Nationalsozialismus' in Kersting, Teppe and Walter (eds) *Nach Hadamar*, pp. 112–36, p. 136.

[18] K. P. Kisker (1976) *Mit den Augen eines Psychiaters* (Stuttgart: Enke), p. iv.

[19] J. Bopp (1980) *Antipsychiatrie. Theorien, Therapien, Politik* (Bonn: Syndikat), pp. 43–44.

[20] S. Thoma (2012) 'Phänomenologisch-anthropologische Sozialpsychiatrie—Wegmarken für eine Theoretische Wiederbelebung', *Psychiatrische Praxis*, 39, 407–409, p. 408.

[21] K. P. Kisker (1960) *Der Erlebniswandel des Schizophrenen. Ein Psychopathologischer Beitrag zur Psychonomie Schizophrener Grundsituationen* (Berlin: Springer), pp. 4–12.

[22] A. Moldzio (2004) *Schizophrenie: Eine Philosophische Erkrankung?* (Würzburg: Königshausen & Neumann), p. 56.

any contact between Kisker and Laing. Later on, Kisker criticised Laing's therapeutic experiment at Kingsley Hall as a 'hippie community' with an atmosphere of 'tolerance and apathy', but at the same time he recognised Laing as 'always more creative and more inventive than his seemingly one-sided and fanatic comrade-in-arms' David Cooper.[23]

In 1963, Kisker became a senior physician at Heidelberg University. With his colleagues Walter von Baeyer and Heinz Häfner, the latter a senior physician who would become head of the department for social psychiatry at the university a few years later, they initiated the most influential historical starting point for later developments towards reform of the mental health care system in Germany. It should be emphasised once again that the engagement of these physicians with the social situation of individuals with mental health problems cannot be understood without their awareness of the crimes in psychiatry under the so-called Third Reich. This fact was emphasised in Baeyer, Kisker—who had the idea for the 'unambiguous' main title[24]—and Häfner's internationally acclaimed 1964 publication *Psychiatrie der Verfolgten: Psychopathologische und Gutachtliche Erfahrungen an Opfern der Nationalsozialistischen Verfolgung und vergleichbarer Extrembelastungen* (*Psychiatry of the Persecuted: Psychopathological and Assessment-related Experiences with Victims of Nazi Persecution and Similar Extreme Stress*).[25] It promoted a re-evaluation of trauma suffered by victims of Nazi violence and thus stood in stark contrast to the anti-victim stance adopted by most German psychiatrists at the time.[26] This anti-victim stance was mainly based on the opinion that traumatised survivors of Nazi violence must have had a 'weak' constitution in the first place and that therefore their mental health problems were not the result

[23] K. P. Kisker (1978) 'Antipsychiatrie' in K. P. Kisker and J. E. Meyer (eds) *Psychiatrie der Gegenwart. Forschung und Praxis, Bd.1 Teil 2. Grundlagen und Methoden der Psychiatrie* (Berlin: Springer), pp. 811–26, pp. 815 and 819. On the person and work of R. D. Laing, see A. Beveridge (2011) *Portrait of the Psychiatrist as Young Man: The Early Writing and Work of R. D. Laing, 1927–1960* (Oxford: Oxford University Press).

[24] Baeyer, 'Baeyer', p. 28.

[25] F. W. Kersting (2007) 'Der Lange Schatten des NS-Krankenmords. Psychiatriereform und "Vergangenheitsbewältigung" in der Bundesrepublik 1955–1975' in S. Oehler-Klein and V. Roelcke (eds) *Vergangenheitspolitik in der Universitären Medizin nach 1945. Institutionelle und Individuelle Strategien im Umgang mit dem Nationalsozialismus* (Stuttgart: Steiner), pp. 361–86, on pp. 377–79.

[26] W. U. Eckhart (2012) '"Entgegen der Stimmung im Eigenen Land …". Deutsche Diskurse um Gewalttraumatisierung im Nationalsozialismus und die "Entschädigung" der Opfer', *Trauma und Gewalt*, 1, 6–15.

of their experience as victims, but of their personal condition.[27] Therefore, German psychiatrists usually did not support compensation for survivors of Nazi violence with mental health problems. This practice was hinted at by the Austro-American psychoanalyst Kurt R. Eissler, who in 1963 posed a provocative question to German psychiatrists: 'The murder of how many of his children must a man be able to stand without symptoms to have a normal constitution?'[28]

It can be assumed that one reason for the initial acceptance of Baeyer, Kisker and Häfner's approach was their reference to the results of international research in the Netherlands, France, Israel and above all the United States.[29] In the US, émigré German psychoanalyst William G. Niederland worked on the same topic and published his influential work on the 'survivor syndrome'.[30] His concept was generally contested by German psychiatrists, and Baeyer was among the very few German colleagues who supported him.[31] Kisker later stated that for him, the positive influence of *Psychiatrie der Verfolgten* on the development of the German compulsory law for Nazi victims was his 'very personal reparation for the disfranchised and injured'.[32]

The practical outcome of these innovative positions was the reopening of one ward of the department of psychiatry clinic in Heidelberg in 1959, the establishment of two rehabilitation wards in 1960 and the implementation of the first night clinic places by Kisker in 1962.[33] Furthermore, the Heidelberg clinic founded a pilot project in 1963 for the professional training of nurses in social psychiatry. The conflict between the nursing staff's traditional attitudes towards psychiatric patients and new forms of social psychiatry such as the therapeutic community was an important

[27] C. Pross (1988) *Wiedergutmachung. Der Kleinkrieg gegen die Opfer* (Frankfurt a.M.: Athenaeum).

[28] K. R. Eissler (1963) 'Die Ermordung von wievielen seiner Kinder Muss ein Mensch Symptomfrei Ertragen Können, um eine Normale Kostitution zu Haben?', *Psyche*, 5, 241–91.

[29] S. Goltermann (2011) *Die Gesellschaft der Überlebenden. Deutsche Kriegsheimkehrer und ihre Gewalterfahrungen im Zweiten Weltkrieg* (München: Pantheon), pp. 317–19.

[30] W. G. Niederland (1964) 'The Survivor Syndrome' in Proceedings of the World Federation for Mental Health, p. 131.

[31] W. Focke (1992) *William G. Niederland: Psychiater der Verfolgten. Seine Zeit – sein Leben – sein Werk* (Würzburg: Königshausen & Neumann), p. 224.

[32] Machleidt, 'Kisker', p. 366.

[33] H. Häfner and H. Martini (2011) *Das Zentralinstitut für Seelische Gesundheit. Gründungsgeschichte und Gegenwart* (München: Beck), p. 32.

problem in those years. In a short study about this phenomenon at the department of psychiatry of Heidelberg University published in 1968, Rolf D. Hemprich and Kisker identified the nurses as the true 'masters of the clinic': in ward meetings the patients had been afraid of criticising their daily life in the clinic because they feared retaliation from the nurses—a result of the 'controlled submissive and latent aggressive' atmosphere that dominated life at the ward up until a few years before. On the other hand, the nurses feared the 'soft wave' of social psychiatry and their loss of power over the patients.[34]

At the same time, Kisker was engaged with attempts to establish social psychiatry in the German Democratic Republic. He helped to organise an 'international symposium on psychiatric rehabilitation' in Rodewisch, Saxony, which was attended by participants from the Soviet Union and other socialist countries, but also from France and Canada. The symposium led to the 'Rodewischer Thesen' (Rodewisch Propositions), demanding an 'active therapeutical attitude' with possibilities of rehabilitation in the psychiatric hospitals.[35] Kisker visited the Rodewisch clinic several times, attended the opening of its first day clinic and kept close contact with the department of psychiatry at the Karl Marx University in Leipzig.[36] This was quite unusual and information about Kisker's motivation for this cooperation remains still remains a mystery.

With this background of experience and progressive engagement, Karl Peter Kisker moved to Hannover in 1966 and became the first chair in psychiatry at the newly founded Hannover Medical School. His inaugural lecture, entitled 'Die Verrücktheit, die Armut und wir' ('Madness, Poverty and Ourselves'), focused on the social causes of the relationship between poverty, madness and deprivation. He referred to the 1961 report of

[34] R. D. Hemprich, and K. P. Kisker (1968) '"Die Herren der Klinik" und die Patienten. Erfahrungen aus der Teilnehmend-verdeckten Beobachtung einer Psychiatrischen Station', *Nervenarzt*, 10, 433–41.

[35] E. Kumbier, K. Haack and H. Steinberg (2013) '50 Years Rodewisch Theses—For the Beginnings of Social-Psychiatric Reforms in East Germany (GDR)', *Psychiatrische Praxis*, 40, 313–20, p. 315.

[36] G. Ehle (2011) 'Psychotherapie in der Psychiatrie' in M. Geyer (ed.) *Psychotherapie in Ostdeutschland. Geschichte und Geschichten 1945–1995* (Göttingen: Vandenhoeck & Ruprecht), pp. 586–90, p. 588; K. Weise (2006) 'Ohne Titel, ohne Kittel' in T. R. Müller and B. Mitzscherlich (eds) *Psychiatrie in der DDR. Erzählungen von Zeitzeugen* (Frankfurt a.M.: Mabuse), pp. 145–52, on pp. 146–47; J. Schulz (2003) 'Die Rodewischer Thesen von 1963—ein Versuch zur Reform der DDR-Psychiatrie' in F. W. Kersting (ed.) *Psychiatriereform als Gesellschaftsreform*, pp. 87–100, p. 89.

the Joint Commission on Mental Illness and Health, the basis of John F. Kennedy's 1963 special address 'Mental Illness and Mental Retardation', and also to studies from the United States and Great Britain dealing with the problem that poor people with mental problems had lesser chances of receiving adequate medical care. Kennedy's special address denounced the 'inadequate conditions in large overcrowded state hospitals' and demanded 'major investment in treatment services at the federal, state and local level' with a main focus on community health centres and outpatient services.[37] Already in 1964, Kisker referred to Kennedy's special address in a programmatic article and stated that 'the psychiatric colleagues and administrators in charge in the USA recognise (easier than ourselves) their own backwardness and now soberly tackle the necessary'.[38] In his view, political and legislative processes in America demonstrated the ability to bring the poor and insane back into the centre of society.

In Germany, as Kisker stated in his inaugural lecture, psychiatric patients were getting 'a pseudo-human minimum care', and its 'meagreness' could only be ignored as long as one was not forced to send one's relative to a big federal state hospital, and was instead financially able to send him or her to a private clinic.[39] Furthermore, his emphasis on the positive developments in the United States and Great Britain put the central problems of mental health care in Germany in focus. Kisker specified them in his 1966 article 'Forderungen der Sozialpsychiatrie' ('Demands of Social Psychiatry'): the geographic isolation of the great psychiatric hospitals, matched by the social isolation of their inmates; the barriers put up by psychiatric practitioners, preventing psychiatric institutions from providing professional help outside the hospital walls; and the immense 'therapeutic vacuum' between hospitals and psychiatric practitioners. To fill this gap, Kisker demanded sheltered workshops for psychiatric patients, night clinics, outpatient clinics, and small psychiatric units in hospitals. He stressed that as long as these therapeutic facilities were not put in place, the new advantages of psychiatric medication 'would go up in smoke'.[40]

[37] M. Halliwell (2013) *Therapeutic Revolutions. Medicine, Psychiatry and American Culture 1945–1970* (New Brunswick, NJ: Rutgers University Press), pp. 202–203.

[38] K. P. Kisker (1964) 'Klinische und Gemeinschaftsnahe Psychiatrische Behandlungszentren Heute und Morgen. Heute in England, Canada und Skandinavien, Morgen in den USA, Übermorgen bei Uns?', *Nervenarzt*, 6, 233–37, p. 234.

[39] K. P. Kisker (2007) 'Die Verrücktheit, die Armut und Wir' in Machleidt, Passie and Spazier (eds), *PsychiaterSein*, pp. 192–203, p. 194.

[40] K. P. Kisker (2007) 'Forderungen der Sozialpsychiatrie', in Machleidt, Passie and Spazier (eds) *PsychiaterSein*, pp. 204–10, on pp. 204 and 208.

These demands were put into practice in Hannover starting with a unit for the further development of deinstitutionalisation in the region, later called by Kisker a 'training camp'.[41] As the buildings of the medical school for the department of psychiatry in Hannover had not yet been finished, the psychiatric clinic of the department (hereafter, the clinic) opened temporarily in 1967 in a building of the mental hospital in Wunstorf, located about 30 kilometres from Hannover's city centre.

The guiding idea of the new clinic was to establish a 'centre for rehabilitation' and a 'model institution to test socio-therapeutic measures'. It contained three wards for 36 patients. Its purpose was to gain experience in developing therapeutic methods for 'the re-integration of mentally ill that had been hospitalised for months and years', and to find therapeutic possibilities for dealing with 'frequent relapses of disease'.[42]

The first patients in this new institution were so-called 'long-term patients' of the Wunstorf hospital. In the new clinic, they were to become part of the therapeutic community concept. The staff consisted of three physicians, one social worker and 15 nurses. The fact that this unit was located in the midst of the Wunstorf hospital dealing with all the problems institutional psychiatry in Germany faced at that time led, however, to permanent conflicts: in the clinic the personnel/patient ratio was nearly one to one, while the Wunstorf hospital had to deal with a physician/patient ratio of one to 100 and a nurse/patient ratio of one to five.[43] On night shifts, for example, one physician in the clinic only had to care for 35 patients, while at the same time one physician of the Wunstorf hospital had to care for more than 1,000 patients. Furthermore, the Wunstorf hospital had the sole duty of the admission of patients in a catchment area containing 1.5 million people.[44] In contrast, the clinic was able to choose its patients.

[41] K. P. Kisker and E. Wulff (1985) 'Psychiatrie' in Rektor der Medizinischen Hochschule Hannover (ed.) *Medizinische Hochschule Hannover 1965–1985* (Hannover: Medizinische Hochschule), pp. 372–80, p. 373.

[42] G. Hofer (1967) 'Aufbau und Gestalt der psychiatrischen Klinik', *Niedersächsisches Ärzteblatt*, 11, 360–62, p. 361.

[43] H. Wittrock (2005) *Niedersächsisches Landeskrankenhaus Wunstorf. Von der Korrektionsanstalt zum modernen Fachkrankenhaus (1880–2005)* (Wunstorf: Niedersächsisches Landeskrankenhaus), p. 132.

[44] M. Bauer (1977) *Sektorisierte Psychiatrie im Rahmen einer Universitätsklinik. Anspruch, Wirklichkeit und Praktische Erfahrungen* (Stuttgart: Enke), p. 16; Wittrock, *Niedersächsisches Landeskrankenhaus Wunstorf*, p. 132.

With the concept of the therapeutic community adapted from Maxwell Jones, the clinic thus increasingly demarcated itself from Wunstorf hospital.[45] By the end of 1967, the clinic took almost no patients from the hospital. As a result of the dynamics of the therapeutic community, the clinic preferred patients that were able to adapt to its newly established methods, and these were predominantly young, intelligent and well-educated. The methods of the therapeutic community in the clinic consisted of, amongst others, group therapy; self-management of the patients (for example by a 'patient committee'); 'educational' presentations by 'interest groups' of patients dealing with 'politics and contemporary history'; 'literature' and 'everyday life psychology'; and a patients' newspaper.[46] Manfred Bauer, senior physician at the department of psychiatry at the Hannover Medical School from 1968 until 1981 and former student of Kisker in Heidelberg, stated in 1977 that three years after its implementation in the grounds of Wunstorf hospital, the clinic had reached an '"Upper Cottage" situation' like the Dutch psychiatrist Jan Foudraine described for his sociotherapeutic model ward in the US clinic Chestnut Lodge.[47] In the words of one local historian, the clinic became a 'noble psychiatric ghetto', with its progressive social psychiatrists looking disapprovingly at the personnel of Wunstorf hospital, while at the same time this personnel regarded the conditions of the clinic with envy.[48] And, as the new head of Wunstorf hospital Asmus Finzen stressed in the 1980s, there was 'no doubt that the drive of social psychiatry' in this situation 'missed out the patients which needed it the most'.[49]

This experimental period is summarised in the following description by the senior physician Helmut Krüger who became head of the clinic in 1970:

> My first impression was (...) to be confronted with a crowd of noisy, scruffy, coffee drinking, cigarette begging noble deadbeats lacking distance and respect for the therapist, and one could be inclined to ignore that they

[45] Bauer, *Sektorisierte Psychiatrie*, p. 16.

[46] G. Hofer (1967) 'Aufbau und Gestalt der Psychiatrischen Klinik', *Niedersächsisches Ärzteblatt*, 7; G. Hofer (1967) 'Aufbau und Gestalt der Psychiatrischen Klinik', *Niedersächsisches Ärzteblatt*, 8.

[47] Bauer, *Sektorisierte Psychiatrie*, p. 17.

[48] Wittrock, *Niedersächsisches Landeskrankenhaus Wunstorf*, p. 132.

[49] A. Finzen (1985) *Das Ende der Anstalt. Vom Mühsamen Alltag der Reformpsychiatrie* (Bonn: Psychiatrie-Verlag), p. 55.

> were patients. (…) It seemed to be a group (…), interacting by rivalling, terrorising and forming cliques. The rehabilitative efforts only led to a permanent 'acting out' (…) themed: 'Just scream it out of yourself'. The team of therapists extremely doubted the possibility of integrating psychiatric patients in this morbid capitalist society. Numerous discussions revealed the will to overcome the psychiatric institution itself, which could only be achieved by overcoming the societal conditions. (…) the team hid itself away on a therapeutic island. This therapeutic island in Wunstorf faced the danger of deforming into an aberrant subculture. The idea of therapeutic community had become an ideology.[50]

In 1972, the clinic finally moved from Wunstorf to the central campus of Hannover Medical School and increased to 117 places and four times as many therapists. This was a crucial break and marked the end of the experimental phase of social psychiatry in Hannover. Now the clinic was re-transformed—like Krüger described it—into a therapeutic institution and it stepped into a 'realistic policy' of social psychiatry.[51]

At the same time as the clinic's relocation, the decision had been made to also change the concept: the aim now was not a therapeutic community for a few patients, but 'community care' for all people in the region. The main reasons for this change lay in the limitations and problems of the therapeutic community itself. Kisker wrote that the experimental phase of the clinic in Wunstorf had emphasised 'rational explicable' and 'psycho-social' aspects of mental illnesses too much and therefore 'banged its head against the wall of madness'.[52] Two of his colleagues later concluded that the 'unavoidable' discovery that patients with schizophrenia could become chronically ill even when they were treated under optimised conditions had been the 'end of great vision' and the first phase of social psychiatry in Hannover.[53]

A team of five psychiatrists, among them Manfred Bauer, had been working since 1970 on a concept of 'sectorised community psychiatric service'—

[50] H. Krüger (1987) 'Reifungskrisen einer Klinik: Anti-institutionelles Wollen und Therapeutische Gemeinschaft' in H. Haselbeck, W. Machleidt, H. Stoffels and D. Trostdorf (eds) *Psychiatrie in Hannover. Strukturwandel und Therapeutische Praxis in einem Gemeindenahen Versorgungssystem* (Stuttgart: Enke), pp. 16–23, p. 17.

[51] Krüger, 'Reifungskrisen', p. 19.

[52] K. P. Kisker (1977) 'Vorwort' in Bauer, *Sektorisierte Psychiatrie*, p. x.

[53] W. Machleidt and T. Passie (2007) 'Einführung' in Machleidt, Passie and Spazier (eds) *PsychiaterSein*, pp. 23–30, p. 27.

for Kisker the solution of the 'institutional crisis' of the clinic at this time.[54] Kisker mentioned 15 years later that the step from the early therapeutic community in Wunstorf to the 'sectorised care' in Hannover had been a 'beneficial disillusionment': the therapist's work had become more 'practical (...) and—as we hope—more effective'.[55] Bauer himself spent a five-month study visit at the London Maudsley Hospital in 1971 and came back inspired by his experiences with complex psychiatric care systems in a community.[56] In cooperation with the psychiatric hospital in Hannover-Langenhagen, the private psychiatric hospital at Ilten and Wunstorf hospital, in 1972 the psychiatric clinic of the Hannover Medical School became the first psychiatric department of a university to take responsibility for one of four regional sectors, including the duty for admission of people in its sector. The system contained social psychiatric services and outpatient clinics, homes for people with permanent mental handicaps, sheltered workshops and a 'transition ward' for persons with enduring mental health problems. By establishing this system, the Hannover Medical School was exceptional in Germany, and the so-called 'Model Hannover' was the first area-wide concept for sectorised psychiatric care in a large German city. For some university clinics it was, however, as Manfred Bauer described in retrospect, a 'provocation':[57] none of them had the commitment to provide therapy in hospital for all psychiatric patients in a defined area as described above—the psychiatric departments of university clinics could instead choose their patients, preferring those with acute illnesses and referring chronic patients to the mental hospitals.[58]

The division of the city of Hannover into sectors took place in a pragmatic way. By the end of the process in 1979, the Medical School's department of psychiatry had to provide care for about 15,000 inhabitants of its sector, or approximately 27 per cent of all Hannover's inhabitants. The main problems in establishing this system were to convince health insurance companies to finance social psychiatric services, the limitation of the

[54] Kisker, 'Vorwort', p. xi.

[55] Kisker and Wulff, 'Psychiatrie', p. 373.

[56] M. Bauer (1997) 'Woher Wir Kommen, Wo Wir Stehen, Wohin Wir Gehen (Sollen). Zur Entwicklung der Sozialpsychiatrie in der Bundesrepublik' in U. Hoffmann-Richter, H. Haselbeck and R. Engfer (eds) *Sozialpsychiatrie vor der Enquête* (Bonn: Psychiatrie-Verlag), pp. 109–21, pp. 114 and 116.

[57] Bauer, 'Entwicklungslinien', p. 6.

[58] Bauer, *Sektorisierte Psychiatrie*, p. 1.

right to choose one's doctor, and the fear of the practitioners themselves of losing their patients to outreach psychiatric services.[59]

Nevertheless, by the early 1970s in Germany the 'inhuman conditions' of psychiatric care were on the political agenda, highlighted by the socio-cultural changes triggered by the student revolts against the system since 1967 and, in Germany, the disputes of a younger generation with their parents, teachers, lawyers, doctors, professors and so on, an older generation that never talked about their role in Second World War, some of whom had escaped a judicial process to examine their activities as members of the Nazi regime.

In this period, important books criticising psychiatry were published in the German language: Michel Foucault's *Madness and Society* (1969), Franco Basaglia's *L'istituzione negata* (1971), R. D. Laing's *The Divided Self* (1972) and Erving Goffman's *Asylums* (1973). On this basis, press articles, television and book reports dealing with the situation in German psychiatric hospitals got public attention. One of the most famous works was Frank Fischer's 1969 book *Irrenhäuser. Kranke klagen an* (*Asylums. The Sick do Accuse*). Fischer was a teacher who had worked for eight months in five psychiatric hospitals as a nurse in order to write his book. His impressions of the humiliation and brutality of everyday life in these institutions led him to compare psychiatric hospitals with concentration camps.[60] Through these developments, 'madness' became a public theme and placed the situation of psychiatric patients onto the political agenda.

In 1970, young psychiatrists formed the Deutsche Gesellschaft für Soziale Psychiatrie (German Society for Social Psychiatry). One year later the Aktion Psychisch Kranke (APK - Operation for People with Mental Illnesses) gathered together politicians of the Bundestag (lower house of the German parliament) and psychiatrists, among them Kisker, to improve the situation of psychiatric patients—but clearly denied to accept them as members of the APK.[61] In 1971, a committee of inquiry was commissioned by the Bundestag to assess the situation of psychiatric care in Germany. The committee consisted of 27 members; all of them—except for one social worker and one psychologist—were practitioners.[62] Its first

[59] M. Bauer (1987) 'Entwicklungslinien der sektorisierten Psychiatrie in Hannover' in Haselbeck, Machleidt, Stoffels and Trostdorf, *Psychiatrie in Hannover*, pp. 2–7, on pp. 4–5; Bauer, *Sektorisierte Psychiatrie*, p. 19.

[60] Brink, *Grenzen*, pp. 444–51.

[61] Brink, *Grenzen*, pp. 471–72.

[62] Brink, *Grenzen*, p. 468.

interim report in 1973 described the 'brutal reality' in the hospitals, with 59 per cent of patients admitted for two years or longer, and 39 per cent in rooms with 11 or more beds.[63] Despite the fact that the report urged the need to downsize psychiatric hospitals, these institutions nevertheless remained at the heart of the concept of community care.[64] The final report of the committee in 1975 also recommended, amongst other things, the establishment and extension of 'community health services', a reduction in size of the large psychiatric hospitals, and the establishment of psychiatric units at general hospitals. It demanded the equal treatment of mentally and physically ill people from legal, financial and social aspects.[65]

The recommendations and demands of the committee's report in 1975 repeated all the main requirements stipulated by the psychiatrists Heinz Häfner, Karl Peter Kisker and Walter von Baeyer in their article 'Dringliche Reformen in der psychiatrischen Krankenhausversorgung der Bundesrepublik' (Urgent Reforms of Psychiatric Patient Care in the Federal Republic) ten years before in 1965.[66] In a way, the demand for 'urgent reforms' was the practical outcome of their approach to mental illness revising traditional psychiatric concepts of schizophrenia and traumatisation. For years, their article had received nearly no response and it took until the end of the 1960s before psychiatry began to appear on the political and public agenda.

The committee's report of 1975 replaced the traditional and custodial concept of psychiatric treatment by means of social psychiatry that Kisker and a few other psychiatrists had been promoting since the early 1960s. Nevertheless, the psychiatric hospital remained the 'node' of mental health care and this care was thus still based on the 'solemnly individual centred way of medical treatment'.[67] Such a situation also reflected the international influences on the reforms of psychiatry in West Germany in the 1970s. British and American models of deinstitutionalisation dominated those reforms. The 'Model Hannover' was clearly influenced by the concept of community mental health services in Great Britain.

[63] H. Häfner (2003) 'Die Inquisiton der Psychisch Kranken Geht Ihrem Ende entgegen. Die Geschichte der Psychiatrie-Enquête und der Psychiatriereform in Deutschland' in F. W. Kersting (ed.) *Psychiatriereform als Gesellschaftsreform*, pp. 113–40, p. 135.

[64] Brink, *Grenzen*, p. 469.

[65] Häfner, 'Die Inquisition', p. 136.

[66] F. W. Kersting (2003) 'Vor Ernst Klee. Die Hypothek der NS-Medizinverbrechen als Reformimpuls' in F. W. Kersting (ed.) *Psychiatriereform als Gesellschaftsreform*, pp. 63–80, p. 72.

[67] Brink, *Grenzen*, p. 473.

Shortly before his retirement in 1987, Kisker confirmed that his department's 'transition ward', established in 1974 for persons with enduring mental health problems, did not necessarily prevent these patients from sometimes being transferred to the still-existing 'long stay wards' in psychiatric hospitals.[68] His admission pointed directly to two crucial problems of the deinstitutionalisation of psychiatry in Germany. First, the existence of psychiatric hospitals far from city centres which had been reduced in size but never closed. And second, the problem of providing adequate community health care for those with enduring mental health problems who did not need therapy but rather a place in which they could live according to their specific needs.

[68] K. P. Kisker (2007) 'MHH-Psychiatrie: Uneingelöstes, Uneinlösbares' in Machleidt, Passie and Spazier (eds) *PsychiaterSein*, pp. 243–46, p. 243.

CHAPTER 6

French Deinstitutionalisation or the Irony of Success: Psychiatrists, the State and the Transformation of the French Psychiatric System, 1945–2010

Nicolas Henckes

The transformations of the French psychiatric system in the second half of twentieth century pose at least two challenges to the concept of deinstitutionalisation. The first lies in the timing and pattern of the decline of psychiatric hospitalisation. Not only was a decrease in the number of psychiatric inpatients observed only after 1967, a relatively late date when compared to trends in countries such as Britain or the United States, but it was widely unanticipated by mental health policies before that date. The second challenge comes from what became of the psychiatric system. If the number of hospitalised patients decreased dramatically after the

N. Henckes (✉)
CNRS/Cermes3, Paris, France
e-mail: nicolas.henckes@cnrs.fr

D. Kritsotaki et al. (eds.), *Deinstitutionalisation and After*,
Mental Health in Historical Perspective,
DOI 10.1007/978-3-319-45360-6_6

1980s, from more than 110,000 at the start of that decade to 56,000 in 2010, the extent to which one may speak of deinstitutionalisation, in the sense of a shrinking of psychiatric services, or of transinstitutionalisation, referring to a transfer of care to other services, remains difficult to assess. By 2000, the cycle of reforms that had affected the mental healthcare system since the 1950s had produced an incredibly fragmented system, of which many parts remained informal and invisible. As a result, what had become of the chronic mental patient was a new, elusive question, a question that has not received an answer to date.

In this chapter I argue that the specificities of the deinstitutionalisation process in France reflected the ways in which mental health policy was negotiated by the psychiatric profession and the state. Until the end of the 1960s discussions over mental health policy were obsessed by what appeared to be an irresistible increase in the number of patients in psychiatric hospitals. For psychiatrists and the state the solution to this problem did not lie primarily outside hospitals. Rather they thought that they had to create a more efficient and humane psychiatric hospital at the centre of a diversified and integrated system that encompassed other services as well. There was little debate on where psychiatric patients had to be cared for, how many psychiatric beds were necessary, where to locate them or over the necessity of developing outpatient facilities in order to better address patients' needs. Debates over mental health policy concerned rather the question of how to plan hospital construction, who would coordinate the system once set up and how patients would circulate through it. The policy that emerged from these debates in 1960, namely sectorisation policy, offered a precise answer to these questions, and hinted that it might produce dramatic changes in the delivery of care for mental patients. Significantly, the launch of sectorisation occurred contemporaneously with the greatest effort in psychiatric hospital construction in France's history.

When the first statistics showing that the number of hospitalised patients had begun to decrease were published in the early 1970s, the perspective that psychiatric patients could be cared for with fewer hospital beds than current standards was beginning to gain ground among parts of the administration and the psychiatric profession. Yet this shift produced little change in the way mental health policy was thought of until 1980. Sectorisation policy had become immensely popular among psychiatrists and administrators, who saw in it a comprehensive solution to the problem of psychiatric chronicity. Advocates of sectorisation took an ambiguous if not ambivalent stance toward the downsizing of hospitals, and sectorisation proved to be a loose framework for fostering change in many parts of the psychiatric system. Although the disability law passed in 1975 proposed the creation of a series of services and benefits for psychiatric patients outside sectorisation, the hostility of the psychiatric profession towards the concept of disability limited their expansion.

The acceleration of deinstitutionalisation after 1980 created a quasi-anarchistic situation. Sectorisation remained the official policy in the mental health field, but it was becoming increasingly at odds both with the reality of healthcare delivery and with planning technologies developed by the Ministry of Health. An increasing number of services and facilities for people suffering from chronic mental illness were created outside sectorisation and escaped state supervision to a large extent. Many patients could not find a way through this maze and ended up in other segments of the institutional treatment of deviance.

This chapter offers a survey of this turbulent story. The first section examines the origins of sectorisation in the debates over the reform of psychiatric hospitals in the immediate post-war period. The second section examines the role of construction standards in framing thinking about institutional settings of mental healthcare delivery from the early 1950s to the 1970s. The last section analyses the conflicts over the definition of care for people suffering from chronic psychiatric illness after the promulgation of the 1975 disability law and the consequences of these conflicts for deinstitutionalisation.[1]

The Politics of Institutional Reform

In the immediate post-Second World War period, the idea that the psychiatric system had to be reformed achieved wide consensus among both French psychiatrists and officials from the Ministry of Health. The problem had been on the agenda of the French Parliament since the early days of the Third Republic and despite heated, decade-long discussions no legislative change had been achieved by the onset of the war. Numerous

[1] This chapter is based on research undertaken for my PhD thesis: N. Henckes (2007) 'Le Nouveau Monde de la Psychiatrie Française. Les Psychiatres, l'Etat et la Réforme des Hôpitaux Psychiatriques de l'Après-guerre aux Années 1970' (Thèse de sociologie, École des Hautes Études en Sciences Sociales). Accessible at: https://tel.archives-ouvertes.fr/file/index/docid/769780/filename/Henckes_These.pdf. I would like to thank the editors of this volume for both inviting me to participate in it and for the generous comments and corrections.

initiatives at the level of both local and central governments had significantly changed the outlook of the system though. Most significantly, in the interwar period dispensaries as well as so called 'open' or 'free departments' (*services ouverts ou libres*) to care for non-committed patients had been opened by several local governments (*départements*).[2] These efforts had been made on a voluntary basis and they had remained limited however. How to generalise them remained a conspicuous and elusive question.

The end of the war seemed to bring about an ideal set of conditions for giving an answer to this question. The quasi-revolutionary political climate that dominated France in the months following the Liberation and a more general impression that the inability of the Third Republic to modernise France had been responsible for the defeat in 1940 instilled a new reformist spirit at all levels of French society.[3] The replacement of older elites with younger men and women who had often gained recognition from participation in the Resistance also seemed to remove some of the pre-war obstacles to reform. The readiness of the psychiatric profession to participate in the creation of a better organised mental health system, contrasting with the laissez-faire attitude that dominated other segments of the medical profession, turned psychiatry into a key field for partisans of organised medicine. While it did not play a role in awaking public awareness (something that would later be described in apocryphal histories), the premature death of tens of thousands of psychiatric inpatients as a result of food restrictions during the war had created room for new approaches in many hospitals.[4] The introduction of shock therapy at the end of the 1930s and the discovery of group therapy practiced in the US and UK in 1945 also inspired a new therapeutic ethos among psychiatrists. Two things accentuated the sense of urgency that surrounded psychiatric reform. First, that the number of psychiatric inpatients might continue to increase and reach the exceedingly high pre-war levels; and second, a sense that psychiatric hospitals were lagging behind general hospitals as a result of underfunding during the pre-war period.

Despite this consensus, just how to organise reform proved to be a divisive issue. During the summer of 1945, a small group of hospital

[2] G. M. Thomas (2004) 'Open Psychiatric Services in Interwar France', *History of Psychiatry*, 15:2, 131–53.

[3] A. Kaspi (2004) *La Libération de la France. Juin 1944–Janvier 1946* (Paris: Perrin); S. Hoffmann (1974) *Essais sur la France. Déclin ou Renouveau?* (Paris: Editions du Seuil).

[4] I. von Bueltzingsloewen (2007) *L'Hécatombe des Fous. La Famine dans les Hôpitaux Psychiatriques Français sous l'Occupation* (Paris: Aubier).

psychiatrists[5] who had been involved in the Resistance organised a meeting in Paris that gathered a large number of psychiatrists in all positions from all over France. The meeting, which was later described by its organisers as the first step in a coming psychiatric revolution,[6] resulted in the publication of a bill to reform the psychiatric system as well as several reports and position papers that would frame discussions for the next few years.[7] The bill did not give practical advice on how to organise psychiatric services but it set forth a framework for creating a more open and integrated system in the framework of a national mental health service. In the line of pre-war discussions the project also devoted much room to rethinking admission processes to psychiatric hospitals. As the bill arrived on the desk of the health minister however, he also received a second bill from the influential child psychiatrist Georges Heuyer, who had been unable to participate in the meeting and contested the public health vision behind his colleagues' project.[8] The Ministry of Health was unable to choose between the two bills and within months the changing political climate made a legislative solution impossible. There would be no new attempt at reforming psychiatric legislation during the next 20 years and, from the late 1940s on, the question faced by reformers from all boards was how to achieve the reform of the psychiatric system.

Hospital psychiatrists committed to reform thought they would find a way to foster change in the system on the basis of initiatives at the local level. In the late 1940s several of them began to experiment with new ways of practising psychiatry with the few available means they had, including some funding from the newly created Social Security. The most influential of these experiments were a series of attempts at transforming the atmosphere in wards through occupational therapy and group techniques inspired by British and American group psychotherapy. By the

[5] In what follows hospital psychiatrists will refer to physicians working in public psychiatric hospitals. Until the end of the 1950s, hospital psychiatrists were the only state-employed clinicians. In the immediate post-war period they formed a small group of *c.*220 individuals with a strong identity and a commitment to both their institution and to clinical psychiatry.

[6] G. Daumézon and L. Bonnafé (1946) 'Perspectives de Réforme Psychiatrique en France depuis la Libération' in *Congrès des Médecins Aliénistes et Neurologistes de Langue Française. 44e Session, Genève et Lausanne, 22–27 Juillet 1946* (Paris: Masson).

[7] X. Abély (1945) 'Avant-projet de Loi sur l'Assistance et l'Hospitalisation des Malades Mentaux', *Information Psychiatrique*, 22, 103–11, 61–67, 88–93, 213–20.

[8] Conseil Syndical (1945) 'Les Problèmes Psychiatriques Actuels', *Information Psychiatrique*, 4, 51 sq.

late 1940s these attempts were theorised under the concept of 'institutional psychotherapy', a concept that soon became immensely popular among hospital psychiatrists and well beyond.[9] Other psychiatrists also created outpatient services as an extension of their hospital activities. One important initiative in the Paris region was the creation of a network of mental hygiene centres under the direction of hospital psychiatrists from the late 1940s on.[10] Other noteworthy experiments included therapeutic homes and workshops.[11] By the late 1950s these experiments constituted a reservoir of interesting innovations but they contributed little to a comprehensive reform. And while they did challenge the institutional order of psychiatric hospitals, they did not contest their role in the treatment of psychiatric illness.

The possibility of a political solution began to resurface in 1947 when the Ministry of Health appointed a commission to advise on mental health problems, the Commission for the Study of Mental Health Problems, which would remain famous under the name it received after its reorganisation in 1949, the Commission of Mental Illnesses.[12] The Commission comprised a majority of psychiatrists representing the diverse sensibilities in the profession, as well as representatives from the various administrations interested in psychiatric matters and was presided over by a psychiatrist. Until 1960, the Commission met on average every three months to discuss decrees, projects from the Ministry of Health or general questions put on the agenda by the Ministry or by psychiatrists themselves. Over the years, discussions at the Commission inspired a series of measures regarding numerous aspects of the organisation of psychiatric hospitals and psychiatry at large. Its work culminated in the second half of the

[9] G. Daumézon and P. Koechlin (1952) 'La Psychothérapie Institutionnelle Contemporaine', *Anais Portuguese de Psiquiatria*, 4, 271–311.

[10] N. Henckes (2005) 'Réformer la Psychiatrie, Organiser les Pratiques de Secteur. La Construction de la Psychiatrie de Secteur dans "l'Expérience du Treizième Arrondissement"' (Rapport pour le programme Sciences biomédicales, Santé, Société, MiRe/INSERM/CNRS).

[11] For instance: E. Diébolt (1997) *De la Quarantaine au Quarantenaire, Histoire du Foyer de Postcure de l' Élan* (Paris: Editions de l'Elan Retrouvé).

[12] Almost all the minutes of the Commission and most of the reports presented during sessions were published by hospital psychiatrists in their professional journal *L' Information Psychiatrique*. Missing items can be found in the national archives, 19950173–1. The reorganisation of the Commission in 1949 was a consequence of its incorporation in the larger Permanent Council on Social Hygiene, one of the two main advisory councils of the Ministry of Health, established in 1938.

decade with a series of discussions that prepared the promulgation of sectorisation policy in 1960.

Sectorisation emerged as an answer to a very practical question. In the early 1950s the French government had decided to invest massively in the health system through the framework of the new planning infrastructure that had been created by Jean Monnet to reconstruct France in the immediate post-war period.[13] For the administration, the question was how to rationally plan the renovation and construction of the psychiatric system: where, how, and according to which architectural design should hospitals be constructed? To what extent should the psychiatric system also comprise other types of service? And who should make such decisions? The ministerial circular that launched sectorisation policy on 15 March 1960 could be read as a guideline regarding how to address these diverse questions.[14]

At the meetings of the Commission of Mental Illnesses, two factions coalesced around two rather different answers to these questions. The first aligned itself behind the small group of hospital psychiatrists who had organised the 1945 meetings. For them, planning should be organised from below, under the responsibility of a clinical psychiatrist—when possible a psychiatrist working in a public psychiatric hospital—in charge of devising the system.[15] The first step of the process was to create outpatient services, notably consultations, and it was only as a last step that the construction of hospitals should be considered. For their opponents, whose leaders were a small group of university-based psychiatrists, the Ministry of Health had to give clear standards on hospital construction.[16] The first issue to address was the institutionalisation of chronic and incurable patients in hospices in order to create room for the treatment of acute patients in hospitals. Day hospitals and other alternatives to hospitalisation could be envisioned but only at a later stage of the planning process. Both

[13] H. Rousso (ed.) (1986) *De Monnet à Massé. Enjeux Politiques et Objectifs Économiques dans le Cadre des Quatre Premiers Plans (1946–1965). Actes de la Table Ronde Tenue à l'IHTP les 24 et 25 Juin 1983* (Paris: Editions du CNRS); B. Jobert (1981) *Le Social en Plan* (Paris: Editions Ouvrières).

[14] Circulaire du 15 mars 1960 relative au programme d'organisation et d'équipement des départements en matière de lutte contre les maladies mentales.

[15] L. Bonnafé et al. (1957) 'Rapport sur l'Équipement Psychiatrique d'un Territoire Dépourvue de Toute Formation Spécialisée', *Information Psychiatrique*, 33:6, 299–311.

[16] J. Dechaume, unpublished note, Commission on Mental Disorders, meeting of the 19 December 1955. Archives Nationales 19950173-1.

groups also diverged on the meaning of hospitalisation. Hospital psychiatrists thought of the psychiatric hospital as a facility for social rehabilitation. Their opponents insisted on the need to think of hospitals as highly technological institutions in charge of delivering cutting-edge treatments, including the emergent and increasingly numerous chemotherapeutic and biological options.

Ultimately, sectorisation was a compromise between both perspectives. The 1960 circular entailed at once a methodology to plan psychiatric institutions and a rationale to manage the system. It was both a guideline and a policy statement. Local governments had to establish plans on the basis of propositions made by hospital psychiatrists and any other interested parties. The programme had to entail the creation of a series of catchment areas attached to psychiatric hospitals, the so-called sectors (*secteurs*). Each sector should be placed under the responsibility of a unique medico-social team, led by a senior psychiatrist from the hospital to which it was attached, in charge of managing hospital beds as well as community services. Local plans were then to be sent to the Ministry of Health to be reviewed by a commission. They were not constraining but they were supposed to stimulate local initiatives.

The idea of coordinating outpatient facilities and inpatient departments through a unique medical direction was the main innovation of the policy. For psychiatrists it was the crux of what they called 'sector psychiatry', a radically new way of practising psychiatry in sectors. The circular gave a list of outpatient services that could be set up, such as day hospitals, therapeutic workshops and dispensaries. These services were not seen as alternatives to hospitals in the sense that they would make it possible to discharge patients trapped in hospitals, but they were supposed to help ameliorate the circulation of patients in the system and thus contribute to its efficiency. Early treatment in dispensaries and aftercare treatment in therapeutic homes were also supposed to reduce unnecessary hospitalisation. Ultimately an efficient mental health system would prevent chronicity. But as we shall see the creation of an array of non-hospital psychiatric services was not considered a reason for planning a reduction in hospital beds.

The Rise and Fall of Construction Standards

The 1960 circular consecrated two important standards that encapsulated the prominent role attributed to psychiatric hospitals in mental health policies in mid-twentieth century France. The first was an architectural model, namely the cottage hospital; the second, a statistical norm, the ratio of three psychiatric beds per 1,000 population. Both standards had circulated before 1960 but their mention in the circular increased focus on them. In the next few years they would be key ingredients in what was probably the biggest construction effort in the history of French psychiatry. When, by the turn 1970s, the role of psychiatric hospitals in the treatment of psychiatric patients began to be contested, the absence of similar standards for outpatient services left mental health planning without strong standards, giving way to a chaotic deinstitutionalisation process.

The cottage model for psychiatric hospital construction, or in French the *hôpital village*, was an output of discussions in the immediate post-war period although it had its origins in discussions dating back from the early twentieth century.[17] The model meant that hospitals should comprise a series of small constructions gathered around central medical and administrative services. It contrasted with the model for hospital construction in the rest of medicine, namely the 'block hospital' model. The cottage model was coherent with a vision of psychiatric hospitals as a place for social rehabilitation. What had been before the war an architectural metaphor was now taken extremely seriously: hospitals had to be true villages that provided a meaningful social life to patients in order to help them readapt to society. The cottage model implied that constructions needed large grounds and henceforth had to be constructed outside urban centres. Since there was also a consensus that hospitals needed to be accessible from cities, the conclusion was that psychiatric hospitals had to be established in suburban areas and be accessible via public transportation. Hospital psychiatrists also insisted that construction should be small in size, with units accommodating not more than 50 beds, and the whole institution not more than 300 beds, but these recommendations were disputed by hospital directors who argued that small hospitals were economically non-viable.[18] In practice, hospitals built from the 1950s to the 1970s often exceeded this standard.

Another motive behind the discussions over the size of hospitals was the widespread perception that the French psychiatric system lacked sufficient

[17] J. Lauzier et al. (1946) *Au delà de l'Asile d'Aliéné et de l'Hôpital Psychiatrique. Documents de l'Information Psychiatrique, 2* (Paris: Desclée de Brouwer). For a history of the model in France, see P.-L. Laget (2004) 'Naissance et Évolution du Plan Pavillonnaire dans les Asiles d'Aliénés', *Livraisons d'Histoire de l'Architecture*, 7, 51–70.

[18] 'Commission des Maladies Mentales. Séance du 29 juin 1954' (1954), *Information Psychiatrique*, 3, 149–55.

beds to meet needs. This perception was encapsulated in the ratio of three psychiatric beds per 1,000 population as the objective for mental health planning in the 1960 circular. Since the last third of the nineteenth century national statistics had documented the steady increase in the number of hospitalised patients and the phenomenon did not seem to halt. In the interwar period most psychiatric hospitals were facing overcrowding. In the post-war period health authorities set out to gain control of the phenomenon. The newly created psychiatry section at the National Institute of Hygiene endeavoured to explain the steady increase in the number of hospitalised patients with the help of statistical methods.[19] These studies seemed to demonstrate that the biggest contribution to this increase was a rise in the number of cases of reactional psychosis, a phenomenon that was related to the modernisation of France. As a result, a decline in the number of hospitalised patients was not to be anticipated.

The two first national construction plans in 1953 and 1957 were devised with reference to an increased rate calculated on the basis of past experience. The reference to the figure of three beds per 1,000 inhabitants in the 1960 circular demonstrated the will of the Ministry of Health to set clearer objectives to local authorities. The figure was cited after a report by the World Health Organization on psychiatric hospitalisation published in 1953[20] but its strength in the French context came from the way it resonated with the interests of the administration: it was a simple and telling standard that slightly exceeded current statistics—France allocated 2.5 beds per 1,000 inhabitants in 1960—so that it seemed both reasonable and still high. Although French psychiatrists might have played a key role in its inclusion in the WHO report, the fact that the figure was cited by an international document made it an indisputable norm.[21] Comparisons were also made with countries with a higher number of beds such as Switzerland or Nordic countries.

[19] H. Duchêne (1949) 'La Statistique des Hôpitaux Psychiatriques. Travail de l'Institut National d'Hygiène', Unpublished Report at the Commission des maladies mentales, 24 May 1949, Archives Nationales, 19950173–1.

[20] World Health Organization (1953) *The Community Mental Hospital. Third Report of the Expert Committee on Mental Health*, Technical Report Series No. 73 (Geneva: World Health Organization).

[21] N. Henckes (2009) 'Narratives of Change and Reform Processes: Global and Local Transactions in French Psychiatric Hospital Reform after the Second World War', *Social Science & Medicine*, 68, 511–18.

The promotion by health authorities of both the cottage model and the three beds per 1,000 inhabitants statistic testified to their commitment to funding psychiatry at a high level. In 1957 the government indicated that psychiatry should attract 30 per cent of all investments in the health sector. Five years later, the Fourth Plan set an objective of constructing 22,630 hospital beds by 1965 and the next plan reaffirmed the objective of creating 18,000 beds by 1970, while it also planned the renovation of 20,000 beds.[22] Even though actual expenditures and constructions were significantly less than planned, the number of constructions rose sharply.[23] From 1956 to 1972, 30 new psychiatric facilities were opened across the country, totalling 13,400 beds in 19 specialised hospitals and 11 psychiatric departments in general hospitals.[24] In 1972, facilities for 7,290 new beds were still under construction. These constructions enabled the steady increase of hospitalisation numbers during the 1960s. The number of hospitalised patients reached a peak in 1967 totalling 120,000 by the end of that year.[25]

Voices anticipating that psychiatry could do with fewer beds were rare but nonetheless significant in the early 1960s. In 1958 the Ministry of Health launched a survey on the impact of the new neuroleptic chemotherapies on inpatient populations. While its results—which were not published—demonstrated acceleration in the annual number of discharges, they did not suggest a decrease in the number of patients.[26] In the next few years advocates of the psychopharmacological revolution such as Jean Delay and Pierre Deniker, who had first described the neuroleptic effects of chlorpromazine, set out to demonstrate that psychotropic drugs had tremendous effects on psychiatric hospital demography, but these views attracted only limited support in the profession.

[22] Commissariat Général au Plan (1961) *Quatrième Plan de Développement Économique et Social (1962–1965). Rapport Général de la Commission de l'Équipement Sanitaire et Social* (Paris); Commissariat général au plan (1966) *Cinquième Plan, 1966–1970. Rapport Général de la Commission de l'Équipement Sanitaire et Social* (Paris: La Documentation Française).

[23] A survey of psychiatric constructions in the 1960s and 1970s still has to be done.

[24] C. Laurenceau (1972) 'Inventaire des Moyens Psychiatriques Existant en France', *Information Psychiatrique*, 48:10, 975–80.

[25] F. Chapireau (2007) 'Le Recours à l'Hospitalisation Psychiatrique au XXe siècle', *Information Psychiatrique*, 83:7, 563–70.

[26] 'Enquête sur la Consommation des Neuroplégiques dans les Hôpitaux Psychiatriques', August 1958, Archives Nationales, 19950173-2.

More significant, and more consequential for construction standards, were a series of pioneering experiments in sector psychiatry launched in the 1960s. The first, and for decades the most influential of these, was created in Paris 13th arrondissement at the turn of 1960 under the aegis of the Association de Santé Mentale et Lutte contre l'Alcoolisme dans le 13ème arrondissement de Paris (ASM13), with massive support from local authorities, the Ministry of Health and Social Security funds. The founder of ASM13, Philippe Paumelle was a hospital psychiatrist who had stopped working in psychiatric hospitals to work for the office of mental hygiene at the Préfecture de la Seine in 1954. Paumelle counted among the most virulent critiques of psychiatric hospitalisation at the time. In the 13th arrondissement he wanted to create *ex nihilo* an institutional system that would be able to address all the mental health needs of the population without relying on existing psychiatric hospitals.

Paumelle did not envision working without hospitals however. In fact the first facility opened by the ASM13 after the consultation centre was a hospital, constructed in the rural city of Soisy-sur-Seine located 30 km south from Paris. The hospital was a typical exemplar of the cottage model. But Paumelle wanted to manage with fewer beds than current standards. Instead of three beds per 1,000 inhabitants (as stipulated in the 1960 circular), the ASM13 would meet the needs of the 13th arrondissement with only 150 beds for a total of 150,000 inhabitants; that is, exactly 1 bed per 1,000 inhabitants. In Paumelle's view, however, working with this small number was only sustainable if the Association could count on a diversity of outpatient services. The plan published by the Association in 1964 mentioned a total of 726 beds and places (*lits et places*), places referring to positions in day-care facilities, including 120 beds in a therapeutic nursing home, 50 beds in a nursing home for elderly patients, 91 places in day hospitals, 240 places in workshops and 50 places in familial placement, totalling 295 beds; that is, two beds per 1,000 inhabitants.

The hospital opened its first buildings in 1963 and reached full capacity in 1968. Creating services in Paris proved much more complicated both because empty locales were missing or too expensive in the 13th arrondissement and because obtaining authorisation from regional and national health and financial authorities proved arduous given the experimental nature of most of the planned facilities. By 1970 the institutional armamentarium of the ASM13 comprised 429 beds and places including 279 beds in the hospital and a nursing home. ASM13 services experienced a period of acute overcrowding in the early 1970s when the

Association had to take over care for a number of chronic patients from the 13th arrondissement who had been hospitalised in other psychiatric hospitals in the region earlier but the situation came under control by the end of the decade.

Despite its leaders' rhetoric, the financial situation of the ASM13 was much too exceptional for it to constitute a generalisable model. In 1967 local authorities still mentioned the standard of three beds per 1,000 inhabitants when they devised a new scheme for planning sectorisation in the Paris region.[27] The scheme detailed a complex organisation. Since all psychiatric hospitals accessible to the Parisian population were located outside the city with the exception of the Sainte Anne hospital, and since sectorisation policy implied that psychiatric services had to be located in the communities, the plan proposed that each sector would be attached to two hospital departments, one, totalling 50 beds, in the Sainte Anne Hospital, and the other totalling 150 beds located in a suburban institution. Since all sectors were supposed to comprise approximately 70,000 inhabitants, the total number of beds matched exactly the standard of three beds per 1,000 inhabitants. In addition, consultation centres, therapeutic homes and workshops, and day care facilities had to be constructed in the sectors.

The first sector following this model was created in Paris 6th arrondissement in 1968 but it rapidly proved to be a failure. The psychiatrist in charge of the service, Paul Bernard, found it impossible to manage two hospital wards in two different hospitals. Outpatient services had been slow to open so that Bernard felt he had been betrayed by his administration. On the other hand he realised that the psychiatric needs of the 6th arrondissement could be addressed with 75 beds in Sainte Anne. This novel organisation was implemented in early 1969. Another sector was created in 1969 with the same number of beds located outside Paris in the Perray-Vaucluse hospital for the 7th arrondissement. In 1969 a document from the administration of the Paris region proposed a revised standard. Proposals should plan for a combination of beds and places totalling three per 1,000 inhabitants.

By 1970 the cottage model had also become increasingly contested. Several calls for rethinking an architectural design compatible with an

[27] A first scheme had been devised in 1963 but it had to be revised because of a reorganisation of the Parisian local government. The analyses in this paragraph and the next are based on the archive of the Direction de l'Action Sociale, de l'Enfance et de la Santé, Archives de Paris, call number 1513W.

urban setting were published in psychiatric journals in the late 1960s.[28] An influential proposal planned to construct hospitals in buildings that would also comprise housing, offices and shops. Other publications suggested the development of psychiatric services in general hospitals. In 1971 a ministerial circular officially declared that constructions should no longer follow the cottage model and that wards should be opened in general hospitals.[29] Large psychiatric hospitals were no longer an ideal.[30]

The Irony of Success: Deinstitutionalisation and the Limits of Mental Health Policy

Despite the evolutions of local and national policies, deinstitutionalisation was still hardly an issue in public debates by the late 1970s. This changed dramatically when, in 1980, Jacques Barrot, the health minister in the last government of the conservative President Valéry Giscard d'Estaing called for the closure of 40,000 psychiatric beds out of the existing 100,000. Although these figures were not very different from the standards circulating at the time, the ministerial call provoked a fierce reaction among psychiatrists, a reaction that was not entirely explained by the poor popularity of the minister among a profession that clearly leaned to the left.[31] Psychiatrists deplored the fact that the minister was only motivated by financial concerns and, above all, that he had not mentioned sectorisation. Sectorisation, in their view, remained the best way to adapt the psychiatric system to both the progress of the discipline and evolving patients' needs.

The psychiatrists' reactions to the ministerial call reflected the immense popularity sectorisation had acquired among the profession well beyond the small circle of hospital psychiatrists who had initially promoted it. In 1964 an influential report by three prominent reformer psychiatrists

[28] N. Sonolet (1966) 'Un Centre de Santé Mentale. Point de Vue et Proposition d'un Architecte', *Information Psychiatrique*, 6, 527–32; G. Ferrand and J.-P. Roubier (1967) 'L'Hôpital Psychiatrique dans la Cité: Programme d'un Hôpital Psychiatrique Urbain de moins de cent Lits', *Recherches*, special issue 'Programmation architecture et psychiatrie', June 1967, 35–136.

[29] Circular 148, 18 January 1971. See M. Audisio (1980) *La Psychiatrie de Secteur. Une Psychiatrie Militante pour la Santé Mentale* (Toulouse: Privat).

[30] The last psychiatric hospital following the cottage model opened in 1983 in the Paris region however.

[31] J. Ayme (1995) *Chroniques de la Psychiatrie Publique à travers l'Histoire d'un Syndicat* (Ramonville-Saint-Agne: Erès).

argued that sectorisation was the only way to address the needs of people with chronic mental illness and to enable them to function at their highest level.[32] Sectorisation also became a key ingredient in the professional identity of psychiatrists. In 1968, after years of demonstrations and mobilisations, psychiatrists secured separation from neurologists, to whom they had been associated since the 1940s, to constitute an independent medical specialty. Advocates of the reform argued that psychiatry represented a specific way of practising medicine that needed a specific organisation. Since ideas of comprehensive care and proximity were at the core of its philosophy, sectorisation was precisely this organisation.[33]

Sectorisation played no small role in the successes of the discipline in the next few years. The 1968 reform had turned psychiatry into a highly attractive discipline, not the least because it had given sectors an important role in the training of psychiatrists and other mental health professionals. The psychiatric profession grew spectacularly from 1,000 practitioners in 1970 to more than 10,000 in 1990. A new breed of psychiatrists was emerging who conceived of themselves as true entrepreneurs and who, thanks to still empathetic and supportive administrations, developed numerous innovative experiments in sectors, such as crisis centres, various forms of day-care facilities and home treatments. Many of these experimental services became models far beyond psychiatry, and sectorisation itself inspired policies in the delivery of healthcare for a variety of populations, from cancer to AIDS and care for elderly people.

By the 1990s, however, the cost of these successes had become increasingly visible. An influential report to the Minister of Health published in 1992 demonstrated that sectors were the sites of extreme inequalities.[34] More than 800 sectors had been created throughout France and if a few of them comprised a number of outpatient services located in communities, the majority had few alternatives to hospitalisation. Even more worrying, many sectors in rural areas still consisted solely of services in large

[32] L. Le Guillant, L. Bonnafé and H. Mignot (1964) 'Problèmes Posés par la Chronicité sur le Plan des Institutions Psychiatriques. Rapport d'Assistance' in P. Warot (ed.) *Congrès de Psychiatrie et de Neurologie de Langue Française. LXIIe session. Marseille 7–12 Septembre 1964. Comptes Rendus*, vol. 2 (Paris: Masson et Cie éditeurs).

[33] *Livre Blanc de la Psychiatrie Française* (1966–1968), 3 vols (Toulouse: Privat) and C. Brisset (1972) *L'Avenir de la Psychiatrie en France* (Paris: Payot).

[34] G. Massé (1992) *La Psychiatrie Ouverte. Une Dynamique Nouvelle en Santé Mentale. Rapport au Ministre de la Santé et de l'Action Humanitaire* (Rennes: Editions de l'École Nationale de la Santé Publique).

psychiatric hospitals located at a distance from communities and with an extremely long duration of stay. While sectorisation policy was supposed to provide higher levels of integration in the psychiatric system, it was in fact creating inequality. And it had proved unable to transform large psychiatric hospitals.

The reasons for these shortcomings were easily identified. In 1983, after years of discussion, the government decided not to institutionalise a formal structure to administer sectors and to give hospitals financial control over outpatient services created in sectors. While this decision did not much change the situation of sector services at first, it soon proved deleterious. As finances became increasingly constrained, hospital directors became reluctant to spend money in services that, in their view, were not the *raison d'être* of their institution. Nursing homes and workshops created in the framework of sectorisation were particularly vulnerable to this logic. On the other hand, legislation treated outpatient services created in sectors as optional. Except for the creation of consultation centres, no standard was given by the administration. A nomenclature was published in 1986 but it was clearly thought of as a repertoire of possible options, rather than a list of recommended services.[35] The list was supposed to be updated as new experiments created new models, but this never happened.

Several reasons explained why sectorisation policy was not more coercive. The government did not want to commit itself to providing the financial support needed to match the standards it would have imposed. Psychiatrists were also reluctant to see administrations set strong standards regarding the organisation of their work. According to them different populations needed different services and as a result it was impossible to declare universal standards. On the other hand psychiatrists insisted that the role of sectorisation policy was to stimulate the creativity of mental health workers, not to set rigid regulations. It was clear, however, that psychiatrists also sought to preserve their autonomy. Significantly, the main outcome of the debates around the 1992 report was the creation of an agency supposed to audit sector services and provide them with advice in order to reform their structures.[36]

[35] 'Arrêté du 14 Mars 1986 Relatif aux Équipements et Services de Lutte contre les Maladies Mentales Comportant ou non des Possibilités d' Hébergement', *Journal Officiel de la République Française*, 19 March 1986, pp. 4620–4621.

[36] The Mission Nationale d'Appui en Santé Mentale, which operated until 2013.

The shortcomings of sectorisation policy were clearly an impetus for the creation of alternative ways of delivering care to psychiatric patients. This did not happen without difficulty however. In 1975 the government passed a comprehensive disability law that, for the first time, sought to address all the needs of disabled people, from school to employment, through accessibility to public buildings and housing.[37] The law created a series of rights, services and institutions, including nursing homes, workshops and benefits, outside the existing infrastructure of hospitals and social security services. A crucial motive behind the legislation was that many disabled people were cared for in hospitals, and especially psychiatric hospitals, where they did not belong. While the official line was that disability and illness were two related concepts, in effect the law created a divide between health and social services for disabled people. A second law promulgated the same day as the disability law organised planning and funding for disability services.[38] The objective was to create an institutional infrastructure that mirrored the organisation of health institutions. In the end planning was even less efficient than in the field of hospitals.

Psychiatric patients were eligible for some of the new disability services: they could access disability benefit and the disability law also created a new category of nursing homes specifically designed for them. These services were promoted by family associations and the government as the first true alternatives to the psychiatric management of mental illness.[39] By contrast, psychiatrists interpreted the law as an attack against both sectorisation and themselves. If the new disability services were supposed to employ some medical personnel, including psychiatrists, they were not directly under their supervision and, moreover, they were only accessible through a procedure monitored by a special, non-medical commission. By the end of the 1970s psychiatrists managed to overturn most of the measures related to psychiatric patients in the disability law, especially those concerning the creation of nursing homes.[40]

[37] Loi n° 75–534 d'Orientation en Faveur des Handicapés, *Journal Officiel de la République Française*, 1 July 1975, pp. 6596–6604.

[38] Loi n° 75–535 du 30 Juin 1975 Relative aux Institutions Sociales et Médico-sociales, *Journal Officiel de la République Française*, 1 July 1975, pp. 6604–6607.

[39] N. Henckes (2012) 'Entre Maladie et Handicap: Repenser la Critique Psychiatrique de la Loi du 30 Juin 1975 d'Orientation en Faveur des Personnes Handicapées', *ALTER, European Journal of Disability Research*, 6, 242–54.

[40] Ibid.

In subsequent years, however, psychiatrists could not prevent the gradual development of a series of services and institutions targeting their patients through the framework of the 1975 laws. By 2000 these services had become a significant force in the institutional management of people with mental illness, and their integration with sectorisation was a major concern for health authorities. Most of these services resulted from local initiatives and were managed by voluntary and charitable organisations, including family associations, a possibility that was encouraged by the 1975 laws. They were also often accessible to people with a diversity of problems that were not always framed in terms of mental health. Authorities had usually only limited knowledge about their activities and no comprehensive directory was available. As a result their distribution over the territory was even more uneven than that of sector mental health services. Although service planning had been a crucial dimension of mental health policy since 1945, many dimensions of the delivery of mental healthcare now escaped state supervision.

Conclusion

By 2014 the number of psychiatric beds had been reduced to 55,800, including 36,000 in public hospitals.[41] France still counted 88 public psychiatric hospitals totalling 26,700 beds, some of which still occupied buildings that had been constructed in the first half of the nineteenth century, and most of the hospitals constructed over the last two centuries were still in service. But psychiatric hospitals had little in common with the total, desperate institutions they had been just a few decades earlier. Deinstitutionalisation in that sense had been a massive reality.

More than the dramatic transformations of the pattern of psychiatric hospitalisation, however, what has dominated French debates over deinstitutionalisation in the last 70 years has been the coordination and integration of services for people with mental illness in the framework of a comprehensive mental health policy. The objective of both the psychiatric profession and health authorities was to shape integrated pathways to care

[41] Direction de la Recherche, de l'Évaluation et des Statistiques (2014) *Le Panorama des Établissements de Santé*, Collection Etudes et Statistiques (Paris: Ministère des Finances et des Comptes Publiques; Ministère des Affaires Sociales, de la Santé et des Droits des Femmes; Ministère du Travail, de l'Emploi, de la Formation Professionnelle et du Dialogue Social).

that would address the needs of patients at every stage of their illness, in all the dimensions of their lives.

In that respect French mental health policy has been a failure. The cycle of reforms that has affected the mental health world since the 1960s has produced an incredibly complex and fragmented landscape of institutions and services, of which no one today has an overall perspective. In this chapter I have argued that this failure of mental health policy was an ironical outcome of its very success. The early successes of sectorisation policy and its enduring popularity among both mental health professionals and the health administration prevented both groups from anticipating and monitoring alternatives to the psychiatric management of patients. As a result, while for most of the last 60 years mental health policy targeted patients with chronic mental illness, in many ways they have become invisible in a system that is to a large extent unable to acknowledge the very existence of their enduring difficulties.

CHAPTER 7

Integration in a Divided World: Salford Community Mental Health Services 1948–1974

Valerie Harrington

Introduction

> It was a very exciting time ... all the principles that have directed my work for forty years since were learnt in that period in Salford ... all the community care stuff at its best we learnt there by doing it.[1]

In a process whose roots lay in the interwar years, mental health care in early post-war Britain was characterised by a shift from a legal/custodial approach to a treatment model of mental disorder and a corresponding move from hospital to community. Within this shift, local authority (LA) mental health departments played an increasingly important role in providing preventive and aftercare services to an ever-expanding extramural

[1] Interview 1: Tom Fryers, Medical Officer in Salford from 1962, and Medical Officer of Mental Health (MOMH) 1965–1969.

V. Harrington (✉)
Macmillan Cancer Care, Manchester, UK
e-mail: val.harrington@icloud.com

D. Kritsotaki et al. (eds.), *Deinstitutionalisation and After*,
Mental Health in Historical Perspective,
DOI 10.1007/978-3-319-45360-6_7

population. This was clearly articulated in the 1957 Percy Report[2] whose recommendations were broadly enshrined in the 1959 Mental Health Act—a landmark piece of legislation which introduced a new, predominantly informal, voluntary system of treatment and created the legal framework for the development of community care.

Analyses of post-war community care services tend to focus on the broad economic and policy contexts within which they developed. The historiography of deinstitutionalisation includes plenty of discussion about the political and therapeutic processes that drove the community care movement and of the economic, administrative and social factors that frustrated attempts to translate the Percy Report into ground level provision.[3] More detailed accounts of the day-to-day functioning of mental health departments are much rarer and tend to focus on services for 'mental defectives' rather than those with a mental illness.[4] The main exception is Freeman's 1984 paper on the evolution of mental health services in Salford where the author had worked as a consultant psychiatrist since 1961.[5] He thus had first-hand knowledge of the facilities and services he describes. This chapter draws on Freeman's work, alongside oral history interviews and a range of primary sources, including annual reports, memos and contemporaneous publications.[6]

[2] Royal Commission on the Law Relating to Mental Illness and Mental Deficiency 1954–1957 (1957) *Report, Cmnd 169* (London: HMSO), pp. 100–247.

[3] K. Jones (1993) *Asylums and After: A Revised History of the Mental Health Services* (London: Athlone Press), pp. 150–58; J. Busfield (1986) *Managing Madness: Changing Ideas and Practice* (London: Unwin Hyman), pp. 326–46; A. Scull (1984) *Decarceration: Community Treatment and the Deviant. A Radical View*, 2nd edn (Cambridge: Polity Press); S. Goodwin (1997) *Comparative Mental Health Policy: From Institutional to Community Care* (London: Sage Publications), pp. 26–112.

[4] J. Welshman (1999) 'Rhetoric and Reality: Community Care in England and Wales, 1948–74' in P. Bartlett and D. Wright (eds) *Outside the Walls of the Asylum: History of Care in the Community 1750–2000* (London: Athlone Press), pp. 217–20; S. Rolph, J. Walmsley and D. Atkinson (2003) '"A Pair of Stout Shoes and an Umbrella": The Role of the Mental Welfare Officer in Delivering Community Care in East Anglia: 1946–1970', *British Journal of Social Work*, 33:3, 339–59.

[5] H. Freeman (1984) 'Mental Health Services in an English County Borough before 1974', *Medical History*, 28:2, 111–28.

[6] The chapter draws on the findings of a Wellcome-funded research study into the post-war history of mental health services in Manchester and Salford. More information about methodology and sources appears in V. Harrington (2008) *Voices Beyond The Asylum: A Post-War History of Mental Health Services in Manchester and Salford* (PhD dissertation: University of Manchester).

Salford was a rare example of the vision of community care contained within the Percy Report, thus providing a valuable window onto the thinking and practices of the period. But its historical significance and interest arguably lie less in its role as exemplar and more in its exceptional nature. This raises important questions. What made Salford so unusual? How did it succeed in translating policy into practice when others struggled? What particular conditions and forces were at play and precisely how did they operate? These can only be answered by focusing on the minutiae of change; by exploring the actors' day-to-day relationships, experiences and motivations and how these interacted with the broader social, geographical and political contexts within which they were situated. Rolph et al.'s study of mental welfare officers (MWOs) in East Anglia and Gittins' history of Severalls Hospital illustrate how oral history is ideally placed to do this, illuminating areas only hinted at in written sources and capturing the 'spirit' of an institution which—intentionally or otherwise—the documentary evidence often fails to convey.[7]

The case study not only sheds new light on a fascinating episode in the history of mental health services but raises important historiographical issues: about the relationship between the national and the local within the change process; and about the place of small-scale case studies within the broader history of deinstitutionalisation.

Local Authority Mental Health Departments

> The mental welfare officers, most of them ... they didn't have any background, you know, they moved into this job as a job, without much, any real training. And so the function was often to chase schizophrenic patients or paranoid patients ... down the streets, grab them, and get them into the ambulance and into the mental hospital, to confine them and keep them out of endangering anybody else.[8]

Mental health departments—the primary site of community care between 1948 and 1971—came into existence under the National Health Service (NHS) Act of 1946. The act created a tripartite system of healthcare

[7] Rolph, Walmsley and Atkinson, '"A Pair of Stout Shoes"'; D. Gittens (1998) *Madness in its Place: Narratives of Severalls Hospital 1913–1997* (London: Routledge).

[8] Interview 2: Mervyn Susser, MOMH in Salford, 1957–1965. Joint interview with Susser and his wife, Zena Stein, who worked closely with him in the mental health department, primarily as a researcher.

whose three branches—general practice, hospitals and public health—had separate administrative structures. Public health remained the responsibility of the LA but its remit was widened to include mental health. This consisted of two main functions; a statutory duty to effect the 'initial care and removal to hospital of persons who are dealt with under the Lunacy and Mental Health Acts' and the rather more open-ended task of providing 'the preventive, and after-care of all types of patients so far as this is not otherwise provided for'.[9]

This second function presented a considerable challenge. Firstly, there was an almost total lack of direction from above. The wording of the 1946 Act and subsequent guidance was vague; its provisions permissive rather than mandatory; and the lack of any specially earmarked funding meant that authorities had neither the incentive nor support to develop anything other than a skeleton service. The 1959 Mental Health Act was more specific, authorising LAs to provide residential, training and occupational facilities and appoint MWOs. Nevertheless, it lacked the level of detail contained in the Percy Report's recommendations, fell short of stipulating minimum standards of provision and, crucially, failed to incorporate any of the committee's careful deliberations about 'finance … and ways of expediting the expansion of community services'.[10] The Act may have provided a clearer legal framework for community care but, like its 1946 predecessor, ultimately lacked the 'legislative teeth'[11] to provide much meaningful or tangible support for those battling to implement it.

Secondly, the lack of support from central government was exacerbated by the politics of public health and the declining status of public health departments within the new NHS. The tripartite system had taken away many of their functions (including responsibility for the increasingly high status general hospitals), creating an organisation that was fragmented and divided. Within this new set-up the Medical Officers of Health (MOHs), whose relationship with the rest of medicine had always been uneasy, felt demeaned and demoralised, their departments sidelined and devalued. Unsurprisingly they did not embrace their new mental health functions with enthusiasm. Neither were the resources to develop new services forthcoming from their LA paymasters. Public health's loss of status within the

[9] City of Salford (1948) *Annual Report of the Medical Officer of Health* [hereafter Salford MOH Reports], p. 108.

[10] Royal Commission, *Report*, p. 242.

[11] Busfield, *Managing Madness*, p. 346.

NHS was accompanied by a diminution of power within the LAs and a loss of influence over their powerful finance committees. Within an increasingly overstretched health and welfare service, the absence of financial ring-fencing for mental health meant that the new departments, whose functions were neither established nor properly understood, arguably suffered most. In the year 1951–1952, for example, LAs in England and Wales spent only £1.3 million in total on mental health[12] and Jones's study of aftercare in Lancashire reveals how, in 1953, most county boroughs employed only two or three mental health officers to cover populations ranging from 59,000 to 167,000.[13]

Thirdly, the background, skills and attitudes of many of these officers were at odds with the philosophy and culture of aftercare and social casework. The LAs had a long history of involvement in statutory mental health work. Prior to 1948 this had been carried out by public assistance relieving officers, an occupational group whose primary concern was with legal and procedural issues rather than the human and relational dimensions of mental healthcare. Many simply transferred into the new mental health departments—a move which, though expedient in terms of manpower, did little to encourage modern, flexible approaches to care. In the early 1950s, in a pattern that reflected the national picture, ex-Poor Law relieving officers accounted for just over a third of Lancashire's mental health workforce. A further 20 per cent was made up of former clerical staff, whose contribution lay more in their 'thorough knowledge of local government practices and procedures which other workers will take years to acquire' than 'in the sphere of human relations' where they were 'inclined to be rigid'.[14] A significant proportion of the rest were mental nurses who, according to Freeman, 'had a completely different background from the Nightingale tradition of general nursing … they were predominantly male, and more closely related to armed forces medical orderlies'.[15] Officers with either a social work background or experience

[12] Welshman, 'Rhetoric and Reality', p. 208. Total LA expenditure on health and welfare around this period ranged from approximately £42 million (1949–1950) to approximately £57 million (1953–1954) (Ministry of Health (1963) *Health and Welfare: The Development of Community Care, Cmnd 1973* (London: Ministry of Health), p. 1).

[13] K. Jones (1954) 'Problems of Mental After-Care in Lancashire', *Sociological Review*, 2:1, 34–56, p. 40.

[14] Ibid., pp. 40–41.

[15] Freeman, 'Mental Health Services in an English County Borough', p. 116. These figures are echoed in a 1959 survey of 32 MWOs in three county councils. Ten were former relieving

of aftercare were thus few and far between. By the end of the decade less than ten per cent of the national workforce had a social science degree or diploma (the usual qualification for those intending to become social workers) and only 26 fully qualified psychiatric social workers (PSWs) were employed in LA settings[16]—hardly a position of strength from which to challenge the 'macho' culture described so vividly by Rolph et al.,[17] or the prevailing view that 'people were either sane, to be left alone, or insane and "subject to be dealt with"'.[18]

These financial, structural and attitudinal constraints meant that, throughout the 1950s, the focus remained almost exclusively on the statutory services and for most authorities aftercare represented little more than an afterthought. Although the 1959 Mental Health Act undoubtedly led to improvements, many departments continued to offer little more than a very basic service.

1948–1957: Salford Mental Health Department: The First Decade

> A gradual process of enlightenment of public opinion towards those suffering from mental illness is becoming evident. Co-operation and understanding are very slowly but surely taking the place of mistrust and derision ... It is not too much to hope that the Mental Health Service through its important features for prevention, care and after-care of the mentally sick, can claim some share of the credit for this.[19]

Post-war Salford was a small (eight square miles) city in the north-west of England, whose industrial heritage was all too evident in the 'smoke polluted air, more houses to the acre than almost anywhere in England',[20]

officers, nine came from a clerical background and seven had been nurses in a mental hospital (H. Miles, J. Loudon and K. Rawnsley (1960) 'Attitudes and Practices of Mental Welfare Officers', *Public Health,* 76:1, 32–47).

[16] Miles, Loudon and Rawnsley, 'Attitudes and Practices of Mental Welfare Officers', p. 33.

[17] S. Rolph, J. Walmsley and D. Atkinson (2002) '"A Man's Job"?: Gender Issues and the Role of Mental Welfare Officers, 1948–1970', *Oral History,* 30:1, 28–41; Rolph, Walmsley and Atkinson, '"A Pair of Stout Shoes"'. The recollections of their informants were echoed in a number of my interviews.

[18] E. Younghusband (1978) *Social Work in Britain: 1950–1975: A Follow-up Study. Vol. 1* (London: George Allen & Unwin), p. 165.

[19] Salford MOH Reports (1950), p. 107.

[20] Ibid., p. 4.

and in the impoverished health and economic circumstances of many of its 175,000 residents. In the early years there was arguably little to distinguish Salford's mental health department from those around it. It faced many of the problems discussed above and throughout the first decade struggled to recruit appropriately qualified staff. George Mountney recalls how, when he arrived as a trainee in 1955, 'the people who were doing the job were people who'd been either clerical officers, maybe in the department, and a job came available and they took it on. And the bloke in charge was an ex-nurse, mental nurse.'[21] The 'bloke in charge', Mr Hope, fitted the stereotypical image perfectly:

> I think Hope really ... admired and aspired to be tough. He knew the job was tough and he admired courage ... You know, he saw it as a tough job and he was strong enough and brave enough to deal with it.[22]

Much of the staff's time was devoted to statutory duties: for the first five years around two thirds of the 300–500 annual referrals were dealt with by admission to hospital, a figure which remained over 50 per cent until the late 1950s.[23]

The annual reports of 1948–1956 suggest a more mixed picture, however. The department promulgated a broader, more accessible vision of mental health than many of its contemporaries. As the author of the 1951 report proudly declared, 'the four simple words "can I help you" have signified the attitude adopted by all members of our Mental Health Service with results that have more than justified our observance of this simple but vitally important approach'.[24] This rhetoric was supported by signs of progress, albeit modest. In 1952 the department adopted a policy of visiting all hospital discharges within a week; and although aftercare constituted only a small proportion of the workload, the shifting patterns of service use (numbers, sources of referral and types of intervention) suggested that patients, relatives and professionals were starting to approach the service in a preventive rather than emergency capacity.[25] Salford was also one of the few authorities to provide social support facilities. In April

[21] Interview 3: George Mountney, MWO in Salford from 1955 and Chief MWO 1964–1970.

[22] Interview 2: Stein.

[23] Salford MOH Reports (1957), pp. 92–93.

[24] Salford MOH Reports (1951), pp. 99–100.

[25] Salford MOH Reports (1952), pp. 127 and 86.

1950 it opened one of the first therapeutic social clubs in the country and four years later converted a room in Cleveland House, a local Maternal Child and Welfare Centre, into a women's day centre. These early developments were far from sophisticated and neither of the premises ideal. But in the context of 1950s Britain they were innovative and forward-thinking, reflecting a service that was prepared to engage with the social, rather than merely procedural, dimensions of mental health care.

The primary driver of these developments was Salford's MOH, Lance Burn. Burn was an unconventional character, 'regarded as a bit of a buffoon by some people because he was always joking and coming out with "outrageous" ideas'.[26] But beneath this deceptive exterior lay a man of great vision and energy, 'one of the few great men at that time still in public health'.[27] His belief that the personal and environmental aspects of public health were inextricably linked translated into an ambitious programme of initiatives, from a pioneering clean air campaign to a range of support services for children and families.[28] Two key principles underpinned his approach: the need for cooperation and joint working between departments and disciplines; and the value of employing well trained professionals. This spawned a variety of creative staffing arrangements, including a joint appointment scheme with the Department of Preventive and Social Medicine at Manchester University, in which the post-holders combined their university roles with clinical work in Salford.[29]

These principles and beliefs were evident in his approach to mental health. He understood mental illness primarily in social and environmental terms and, while never questioning the need for psychiatrists and mental hospitals, regarded the community as the natural site of mental healthcare. He welcomed his department's new responsibilities and sought to create a service that reflected and fostered the changing attitudes to mental illness and worked closely with existing health and welfare organisations. The extent to which he was able to achieve these aims was, however,

[26] Interview 4: Joyce Leeson, Medical Officer and researcher in Salford 1954–1965.

[27] Interview 1: Fryers. Everyone I interviewed who had known Lance Burn shared Fryers' opinion.

[28] For a brief biography of Burn see W. Elwood and A. Tuxford (1984) *Some Manchester Doctors: A Biographical Collection to mark the 150th Anniversary of the Manchester Medical Society* (Manchester: Manchester University Press), pp. 175–80. For his approach to public health see J. Burn (1959) *Recent Advances in Public Health*, 2nd edn (London: J. & A. Churchill).

[29] Salford MOH Reports (1954), p. 7.

limited. Unusually, the problem was not one of LA funding or support. Early post-war Salford was 'a strong Labour borough with a progressive reputation, a progressive intent, and enormous problems',[30] and Burn's quest to improve the health of the city fitted perfectly with its councillors' strongly socialist agenda. And, as is evident from the language and tone of his annual reports, he was a smooth operator, knowing exactly how to charm and engage his audience. He had an excellent relationship with the Health Committee and Mental Health Sub-Committee and, in contrast to many of his contemporaries, was successful in securing reasonable funding for mental health. Perversely, the main obstacles came from within, in the form of the department's strongly traditional workforce. Burn's attempts to counter this by advertising for appropriately qualified staff proved fruitless, reflecting the national scarcity of PSWs and candidates with social science degrees. Despite some promising developments, and despite the rhetoric of the annual reports, the ethos of the department thus changed little over the period and the staff remained entrenched in a culture of containment and control.

Burn, however, built the foundations for change. These lay in the attitudes and structures of the public health department: its openness to new ideas; its energy in identifying and tackling significant health and social problems; its strong internal and external communication channels; its flexible staffing arrangements; and the goodwill and financial backing of the council. And, despite Burn's limited successes in mental health, the gradual shift towards more preventive services suggests that even the traditionalists were not totally immune to his influence. The actual agent of change, Mervyn Susser, was an outsider—though it was Burn who was responsible for bringing him to Salford; and Burn who, true to form, was prepared to take a risk on a South African epidemiologist who had no experience of psychiatry and had never been near a LA health department.

1957–1961: A Service Transformed

> I had a very different perspective on what a mental health department should be doing … and from my new found knowledge [laughs] … I would try and lead them into another perspective than the capture of people who were crazy and getting them into hospital.[31]

[30] Interview 4: Leeson.

[31] Interview 2: Susser.

In February 1957 Susser took up his new role as part-time Medical Officer of Mental Health (MOMH), one of the joint appointments that Burn had negotiated with Manchester University. Heavily involved in South African anti-racist politics since his student days, Susser had arrived in Britain the previous year to escape an increasingly hostile political situation. On the face of it he was spectacularly unqualified for the post. But what he brought was a set of personal and professional beliefs developed while working as a physician in a community health centre in Alexandra, a black township on the outskirts of Johannesburg.[32] Informed by the emerging discipline of social medicine, he based his approach on two key principles: that healthcare policy should be underpinned by epidemiological data; and that epidemiological data could (and should) never be separated from its social context. Faced with the task of applying these principles to mental health services in Salford, Susser started by 'read[ing] everything I could lay my hands on'. Significantly, this included the Percy Report, which was fortuitously published within weeks of his appointment. With its emphasis on community care and maintaining the social functioning of the mentally ill, Susser recalls how the report offered a vision that chimed with his own philosophy of social medicine—'it just fitted ... the Royal Commission was my model'.[33]

Susser quickly recognised that his team neither shared this vision nor had the skills to realise it. What was needed was a more professional workforce, equipped to both deal with the increasingly complex demands of mental patients and their families, and work effectively with colleagues in other parts of the mental health service. Given the difficulties in recruiting qualified staff, his only option was to professionalise the workforce from within. The main vehicle for this was the weekly case meeting—an institution he imported almost wholesale from Alexandra. Described in the 1960 report as 'a means of training in casework and research, and a channel of communication through which policy can be transmitted from above, or modified from below',[34] the case meeting operated on a number of levels. Its primary purpose was to educate—to inform, but also to encourage a more open, critical and questioning approach. But, as Mountney recalls, it was also 'a way of bringing his influence ... to try and change the balance

[32] M. Susser, Z. Stein, M. Cormack and M. Hathorn (1955) 'Medical Care in a South African Township', *Lancet*, 265, 912–15.

[33] Interview 2: Susser.

[34] Salford MOH Reports (1960), p. 94.

of what we were about'[35]—the central element in Susser's programme of cultural change.

This transformation did not occur seamlessly. Mountney recalls how Hope 'wasn't overjoyed … well, it was a totally alien thing to have this bloke from university, not a psychiatrist or anything, coming in and doing all sorts of odd things, as it were', while other 'old stagers … were glad to get out of it, it was outside their ken'.[36] Those who remained, however, responded with growing enthusiasm. There were a number of reasons for this. Firstly, the seeds of change had already been sown: Susser's approach, though novel, accorded with the foundations and principles already established by Burn. Secondly, aware of the anxieties and insecurities that his changes might provoke, Susser deliberately worked at building up the confidence and self-esteem of his staff. Key to this was the departmental ethos. Although very much the leader, Susser fostered an open, egalitarian feel in both the meetings and the organisation as a whole—an ethos that was to continue after his departure under Mountney's leadership: 'I just remember that they were inclusive meetings, yeah … there was openness to ideas from anywhere';[37] 'a sense that there was more to learn and we were all there together, learning it and sharing it'.[38]

The third factor was the presence of a natural ally: George Mountney. Unlike his colleagues Mountney had a social science diploma and, as he recalls, 'was the one person that, you know, adopted what you might call a more progressive approach to things. And when Mervyn arrived I realised that we could go places, as it were, his ideas and mine.'[39] In the early years Mountney played a pivotal role in bridging the cultural gap between Susser and his workforce:

> I think I played some part in enabling Mervyn to do his stuff ... I was able to mediate, if that's the word, between Mervyn's approach, his research approach, his academic approach, and the troops that were at the front line.[40]

[35] Interview 3: Mountney.

[36] Ibid.

[37] Interview 5: Bill Douglas, Deputy Chief MWO, 1965–1970.

[38] Interview 6: Monica Baynes, MWO in Salford, 1967–1970.

[39] Interview 3: Mountney.

[40] Ibid.

With Susser's encouragement and support, Mountney went on to take a leading role within the department. Following his secondment to the PSW course in 1960 he was appointed social work supervisor, responsible for training and casework supervision, and in 1964 took over the running of the department as Chief MWO. Administrative functions were delegated to a colleague—a division of labour that marked the shift from LA administration to professional service provision.

Reform from within was only part of the picture. On his arrival, Susser embarked on a systematic service analysis. He turned initially to the rather meagre statistics routinely collected by the department: staffing levels, referrals, numbers and types of admissions and the proportion of patients receiving preventive or aftercare services. These had been dutifully recorded in the annual reports, but with little comment or discussion. Susser changed this. The 1957 report stands in stark contrast to its predecessors: the data are more extensive; presented more clearly, using comparative graphs and tables; and analysed in detail, to demonstrate how, over time, 'therapy begins to replace the custodial approach'.[41] The following year he enlarged the data set to include the demographics, referral details, service histories and current contacts of all Salford patients receiving inpatient, outpatient or LA care. His analysis of these data, presented in the 1959 report, revealed a startling lack of coordination between the various branches of Salford's mental health services.

The centrepiece of the report was a diagram (see Fig. 7.1) showing the flow of patients through the services and the connections between the LA, general practitioners (GPs), the main mental hospital (Springfield) and the general hospitals (Salford Royal and Hope), which provided outpatient services and a handful of voluntary admission beds. What is striking is the total lack of contact between the mental hospital and either GPs or outpatient consultants. This partly reflected how, prior to the 1959 Act, LA officers were legally responsible for all mental hospital admissions. But it was also symptomatic of a system in which the component parts functioned almost totally independently of one another—resulting, Susser argued, in damaging discontinuities in patient care.

These discontinuities occurred at all stages of the process. The apparent two-way communication channel between mental hospital and LA amounted to little more than the delivery of a patient to the doorstep of the hospital by the MWO and the posting of a brief note to the mental health department sometime after discharge. MWOs thus had no contact

[41] Salford MOH Reports (1957), p. 91.

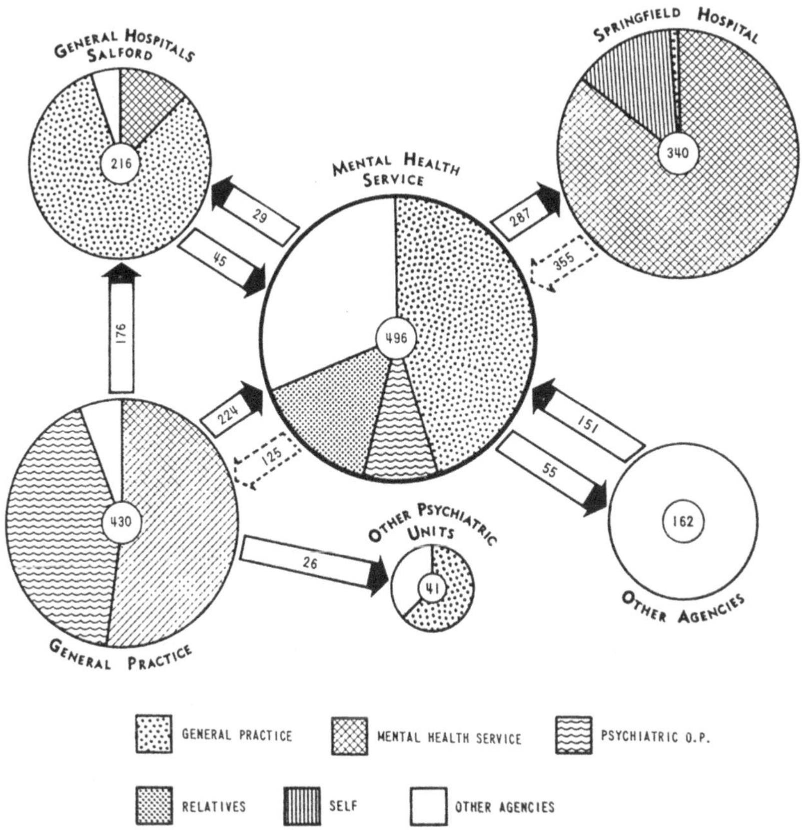

Fig. 7.1 The flow of referrals between agencies (M. W. Susser (1960) *A Report on the Mental Health Services of the City of Salford for the Year, 1959*, City of Salford Health Department, 5, p. 4).

with patients during their hospital stay, no warning of their imminent discharge and virtually no information about the care and treatment they had received. Relationships between GPs and MWOs were also difficult. Susser's analysis of GP referral patterns suggested that a worrying proportion regarded the service as little more than an 'admitting'—as opposed to a 'casework' or 'sorting'—agency.[42] Similarly, outpatient psychiatrists

[42] Salford MOH Reports (1958), p. 99 and appendix II.

rarely referred patients for casework. The department may have seen itself as shifting towards a more preventive and supportive service, but it had failed to communicate this to its key partners.

This fragmentation was largely the product of the tripartite NHS, whose structure discouraged cooperation between its culturally and organisationally distinct branches. Susser could do nothing about the overarching structure—but he could work on improving relationships between the various elements, a process he termed 'active coordination'. In 1957 he established monthly case conferences with hospital staff and GPs. Two years later he reorganised caseloads, basing MWOs around GP practices rather than geographical areas. Each doctor now had an identifiable officer to work with. Despite a mixed response—'some [GPs] would take no notice of them [MWOs], some ignored them deliberately, some worked closely with them. But whether they ignored them or not they had one'[43]—the statistics showed a marked difference in referral patterns, reflecting a much closer working relationship between LA and GP.

Establishing closer relationships with and between the hospital services proved more problematic. Susser's proposal for improving links between the mental hospital and general hospitals by getting psychiatrists to work across both institutions, met with resistance from both sides; while the secondment of an MWO to work in one of the outpatient clinics was abandoned following objections from hospital-based social workers. Staff at Springfield were similarly unwelcoming: attempts to introduce MWOs onto the wards to visit patients and liaise with staff failed; while efforts to establish close working relationships with Dr Blair, Springfield's Medical Superintendent, met with only limited success. Susser invited Blair to become Honorary Consultant to the Mental Health Service and, from 1959, Blair attended the department's weekly meetings. Susser was characteristically diplomatic in his published accounts of these encounters, describing how Blair's attendance 'tended to turn the meeting into a duologue in which the two doctors stated their views, often opposed, on the proper use of services'. While 'this helped to clarify the difficulties in the way of co-ordination',[44] it did little to further the cause of service integra-

[43] Interview 1: Fryers.

[44] M. Susser (1962) 'Changing Roles and Co-ordination in Mental Health Services', *Sociological Review Monographs*, 5:S1, 61–90, pp. 80–81.

tion, and a lengthy piece in the 1960 annual report suggested that stalemate had been reached.[45]

Having exhausted the communication channels identified in his 1959 diagram, Susser responded by creating new linkages in the system. This took the form of a new consultant post spanning the three branches of the service. In 1961 Hugh Freeman, a young psychiatrist from the prestigious Maudsley Hospital, was appointed to work across the mental and general hospitals and for one session a week with the LA. What Susser modestly described as a 'single administrative adjustment'[46] translated into a highly complex and challenging role. But its creation transformed relationships across the service, providing the necessary conditions for the integrated care envisaged by the Percy Commissioners.

Though rare, other examples of joint hospital–LA arrangements did exist, the most cited being Nottingham, Oldham and, slightly later, Croydon.[47] Almost without exception, the initiative for these schemes came from hospital doctors—in contrast to the Salford service whose centre remained firmly in the LA and which, through the 1960s, became increasingly MWO- rather than doctor-led.

1961–1974: Integration and Disintegration

> My main impression of Salford was the ease of communication between the different professionals in a way that I'd never experienced before, and that was enormously to the patients' benefit. We could see them at the early stage of referral, say from the GP, [because] if one had to be involved in compulsory admission then one knew the patient beforehand and one knew the GP, so that was much more straightforward. One knew the consultant very well, and one had easy access to the wards to follow up and similarly when the patient was discharged.[48]

With the twin themes of professional development and service coordination now embedded in the structures and practices of the department,

[45] Salford MOH Reports (1960), pp. 93–95.

[46] Susser, 'Changing Roles', p. 88.

[47] D. Macmillan (1956) 'An Integrated Mental-Health Service: Nottingham's Experience', *Lancet*, 268, 1094–95; H. Freeman (1960) 'Oldham and District Psychiatric Services', *Lancet*, 275, 218–21; A. May and E. Gregory (1963) 'An Experiment in District Psychiatry', *Public Health*, 78, 19–25.

[48] Interview 6: Baynes.

Freeman's appointment marked the beginning of a more settled period—but one that was characterised by almost constant growth. As Freeman remarked, 'there was a feeling of activity that was simply not there in most parts of the country'.[49]

I have described these developments elsewhere.[50] In brief, the number of clients, staff and facilities increased, as did the range and frequency of inter-professional interactions. Salford's growing reputation, reinforced by its close links with the local PSW and social work courses, meant that recruitment problems became a thing of the past. This was not the case in other LAs. In 1969, for example, Manchester blamed a reduction in preventive and aftercare work on the 'continued shortage' of MWOs. The 'impossibility' of recruiting qualified staff had forced them to appoint nine trainees to work alongside their eleven MWOs and two PSWs.[51] In contrast, Salford, whose population was a quarter of Manchester's, had nine officers—now called 'mental health social workers' (MHSWs)—seven of whom had social work qualifications, including five PSWs.[52] This was reflected in the broader and more complex range of responsibilities they took on, and their increasingly central and autonomous position within the service. New facilities, including two hostels, were also opened, and the social club and day centre upgraded to reflect more up-to-date therapeutic approaches. Such facilities remained a relative rarity: by 1973 there were only 97 day centres and 167 hostels for the mentally ill in England.[53]

'Active coordination' continued apace. Freeman expanded facilities at Hope and introduced partial sectorisation into Springfield. All Salford patients now came under his care, providing a more personal service for patients and, for the first time, easy and straightforward communications between psychiatrists and LA staff. These were reflected in a more 'flexible use of available resources, rather than a rigid assessment to one sector of care',[54] with inpatients from Hope attending Cleveland Day Centre; Springfield patients joining the social club; regular visits by hospital psychiatrists to community facilities; and joint MHSW/psychiatrist assessments. The creation, in 1966, of four joint appointments formalised the

[49] Interview 7: Hugh Freeman, consultant psychiatrist in Salford 1961–1988.

[50] Harrington, 'Voices Beyond The Asylum', pp. 102–24.

[51] City of Manchester (1969) *Annual Report of the Medical Officer of Health*, pp. 150, 145.

[52] Salford MOH Reports (1969), p. 138.

[53] Welshman, 'Rhetoric and Reality', p. 214.

[54] H. Freeman (1980) *Mental Health Services in Salford 1948–74* (M.Sc. Dissertation: University of Salford), p. 123.

principles of 'joint working'. The post-holders spent half their time in hospital, working in a PSW capacity, and the remainder in their LA bases. The same year saw the launch of family doctor teams, which brought together GPs, district nurses, midwives, health visitors and MHSWs. The latter spent half a day a week in the surgery, seeing patients, holding discussions with the doctor and participating in team meetings.[55]

Tensions, frustrations and setbacks inevitably accompanied these developments, alongside staff changes, including Susser's departure in 1965. Nevertheless the department remained stable and cohesive, balancing its strong sense of identity and purpose with an eagerness to respond openly and flexibly to new challenges. Its ethos and structures were not enough, however, to protect the service from the massive upheaval it experienced at the end of the decade.

In 1968, the balance of power in Salford council swung from Labour to Conservative. In March 1969, the new chair of the finance committee declared that Salford's health expenditure was 'in some categories excessive and well out of line with national county borough and comparable cities averages ... now is the time for action'.[56] His words clearly applied to the mental health service, whose funding at £700 per 1,000 population was almost twice the national average.[57] Threatened with major budget cuts the department mounted a huge campaign which, though partially successful, marked the end of its special relationship with the health committee. This was sealed by Burn's retirement in the same year and his replacement by a very different breed of MOH. Described by Mountney as 'a bureaucrat down to his socks',[58] Dr Roberts, according to Bill Douglas, Mountney's deputy, seemed to lack the clout and the vision of his predecessor, and be willing to implement the Health Committee's cuts in 'an area he didn't value so highly'.[59] The two events precipitated the departure of most of the MHSW team, including Mountney and Douglas.

A third factor also drove this staff exodus: the transfer, in 1971, of LA mental health functions from public health into the new social services departments. As recommended in the 1968 Seebohm Report, these new departments had a generic structure. This was designed to overcome the

[55] Salford MOH Reports (1966), pp. 140, 150.

[56] 'The Need for Health Cuts', *Salford Reporter*, 28 March 1969.

[57] 'Plan to Close Mental Health Centres', *Guardian*, 31 March 1969, p. 4.

[58] Interview 2: Mountney.

[59] Interview 5: Douglas.

rigid demarcations that had grown up around the four main branches of welfare by creating a single, comprehensive service for children, the elderly and people with mental or physical disabilities. While these aims clearly chimed with the ethos of the mental health department, the reality proved somewhat different. The MHSWs initially retained their specialist functions, but their move into 'patch-based' generic teams meant that they no longer worked together. The support and cohesiveness that had been such a feature of the old service soon dissipated. During the first year ex-department members tried to keep up the old traditions. They continued to hold weekly case conferences, to which they invited their generic social work colleagues, and those holding joint appointments maintained the link between hospital and community. But as the two cultures grew further apart, the gap between the two settings widened. By 1974 referrals had almost halved and short-term crisis intervention was the norm.[60] The reorganisation of LA boundaries in the same year heralded a shift to fully generic social work. Those wishing to pursue a specialist career retreated to hospital-based posts. Ironically, the attempt to provide a more integrated approach to welfare had resulted in the demise of one of the most integrated mental health services in the country.

The picture was similar across the country. Within the new generic departments mental health occupied a lowly position: only two of the 174 social services directors were former Chief MWOs;[61] growing public concern about child abuse meant that childcare became the main priority;[62] and mental health work became increasingly unpopular amongst social workers.[63] Following the 1974 NHS reorganisation, specialist mental health social work became almost totally confined to hospital PSW departments. In 1975, the government White Paper, *Better Services for the Mentally Ill*, criticised the dominance of hospital services and the inadequacy of com-

[60] K. Wooff (1978) *The Use of Social Services by Psychiatric Patients in Salford* (M.Sc. Dissertation: University of Manchester); R. Bhaduri (1976) *A Study of Mental Health Services in Salford from 1960 to 1974* (MA Dissertation, University of Manchester).

[61] J. Smith (1972) 'Top Jobs in the Social Services' in K. Jones (ed.) *The Year Book of Social Policy in Britain 1971* (London: Routledge & Kegan Paul), pp. 16–30.

[62] F. Martin (1984) *Between the Acts: Community Mental Health Services 1959–1983* (London: Nuffield Provincial Hospitals Trust).

[63] J. Neill, W. Warburton and B. McGuiness (1976) 'Post Seebohm Social Services: (1) The Social Worker's Viewpoint', *Social Work Today*, 8:5, 9–14.

munity provision.[64] Whatever the failings of the previous system, the new LA structure had done little to progress the cause of community care.

Conclusion: Narratives of Change: National Policy and Local Context

> Innovations do not all come from the centre, or through extensive policy reviews; some come, rather quietly, from the historical peculiarities of particular people and places.[65]

As John Pickstone points out, the history of the health services tends to be written from a top-down perspective in which policy formation at the centre is followed by policy implementation at the periphery.[66] Clearly, Salford both reflected and was framed by the policies and legislation of the early post-war period. The department grew out of the 1946 NHS Act; the Percy Report shaped Susser's ideas; Salford's service principles and practices were perfectly in tune with both the spirit and substance of the 1959 Act; and the Seebohm reorganisation undoubtedly contributed to its demise. However, as Nicolas Henckes argues, local service developments constitute not 'just variations within a grand narrative'[67]—whose role is primarily to illustrate and exemplify—but two-way interactions in which, I would argue, the 'master narrative' plays an important role, but the precise form and direction is largely dictated by the particularities of the local setting.

Salford, for example, did not simply respond to national directives. The 1959 Act served mainly to support and strengthen measures that the department were already putting in place—measures that were far more sophisticated and wide-reaching than anything contained in the policy documents. Meanwhile the seeds of its destruction lay as much in local politics as national policy. The role of these local factors was arguably

[64] DHSS (1975) *Better Services for the Mentally Ill, Cmnd 6233* (London: HMSO), p. ii.

[65] J. Pickstone (1992) 'Psychiatry in District General Hospitals: History, Contingency and Local Innovation in the Early Years of the National Health Service' in J. Pickstone (ed.) *Medical Innovations in Historical Perspective* (Basingstoke: Macmillan), p. 199.

[66] Ibid., p. 185.

[67] N. Henckes (2009) 'Narratives of Change and Reform Processes: Global and Local Transactions in French Psychiatric Hospital Reform after the Second World War', *Social Science and Medicine*, 68:3, 511–18, p. 511.

amplified by the permissive nature of post-war mental health legislation. This worked both in Salford's favour, enabling it to develop innovative policies; and to its detriment, leaving it vulnerable to collapse when local politics shifted. Central to the narrative is the way in which key actors interacted with these political forces. But equally important was Susser's introduction of team structures and collective working practices that transcended his leadership. This combination of national and local, political and personal forces resulted in a service that is best seen not as an example of 1960s LA mental health provision, but more as an outlier or exception. The 'point' about Salford, therefore, is not that it grew out of or exemplified a 'grand narrative' of community care, but the ways in which its philosophy, practices, politics and personnel interacted with that narrative.[68]

The extent to which Salford influenced national developments in mental health is difficult to assess. Certainly its senior practitioners were prolific writers, publishing a range of academic articles; several team members went on to work in education; and many talked about how their Salford experience profoundly influenced their future practice. Within the broader historiography of mental health care, its main importance lies in what it tells us about the precise processes of, and obstacles to, micro-level change. Clearly, it is necessary to identify and elucidate broad national trends within this history. But it is equally important not just to offer examples of how these played out on the ground—Henckes's 'variations of the grand narrative'—but to work at unravelling the highly complex, contingent and dynamic web of local and national interactions that shaped such variations. Drawing on the activities, experiences and interactions of actors whose voices we rarely hear, a local case study such as Salford is uniquely placed—not merely to exemplify, illustrate and bring to life, but also, to echo Pickstone, to rather quietly inform, extend, challenge and even reframe our relationship with, and interpretation of, the grand historical narratives of deinstitutionalisation.

[68] V. Harrington (2010) 'Learning about Mental Health Services through Local Histories: Case Studies from the Manchester Region' in A. Anderson, W. Hubbard and T. Ryyman (eds) *International and Local Approaches to Health and Health Care* (Oslo: Novus Press, 2010), pp. 75–78.

CHAPTER 8

Initiating Deinstitutionalisation: Early Attempts of Mental Health Care Reform in Greece, 1950s–1970s

Despo Kritsotaki

Introduction

The second half of the twentieth century was a time of reorganisation of mental health services in the Western world, including the downsizing and/or closing down of asylums and their replacement by various models of 'community care'.[1] Between the 1940s and 1980s different countries experienced 'deinstitutionalisation', or the shift from asylum to

[1] The bibliography on deinstitutionalisation has been growing during recent years. See indicatively, J.-C. Coffin (2005) '"Misery" and "Revolution": The Organisation of French Psychiatry, 1900–1980' in M. Gijswijt-Hofstra, H. Oosterhuis, J. Vijselaar and H. Freeman (eds) *Psychiatric Cultures Compared: Psychiatry and Mental Health Care in the Twentieth Century: Comparisons and Approaches* (Amsterdam: Amsterdam University Press), pp. 225–47; H.-P. Schmiedebach and S. Priebe (2004) 'Social psychiatry in Germany in the Twentieth Century: Ideas and Models', *Medical History*, 48, 449–472; and D. MacKinnon and C. Coleborne (2003) 'Introduction: Deinstitutionalisation in Australia and New Zealand', *Health and History*, 5, 1–16.

D. Kritsotaki (✉)
Independent scholar, Athens, Greece
e-mail: despo.kritsotaki@gmail.com

D. Kritsotaki et al. (eds.), *Deinstitutionalisation and After*,
Mental Health in Historical Perspective,
DOI 10.1007/978-3-319-45360-6_8

community, in different ways, times and paces and to different extents. Although deinstitutionalisation continued practices already in place since the late nineteenth century,[2] it should be seen as a novel post-1945 phenomenon: after the Second World War more emphasis was placed on the ambulant sector and outpatient services expanded,[3] while professionals were more than ever committed to the reform of mental healthcare.[4]

In this chapter, I discuss early versions of deinstitutionalisation in Greece. While an official deinstitutionalisation policy was inaugurated only in the 1980s, since the 1950s there were efforts to reorganise mental healthcare, mostly undertaken by private initiatives—mental health associations and private institutions. These initiatives have not been systematically studied and therefore their part in mental healthcare reform remains unclear.[5] They are both seen as marginal initiatives which did not challenge established

[2] M.-C. Thifault and I. Perreault (2012) 'The Social Integration of the Mentally Ill in Quebec Prior to the Bédard Report of 1962', *Canadian Bulletin of Medical History/Bulletin Canadien d'Histoire de la Médecine*, 29, 125–150; M. Thomson (1998) *The Problem of Mental Deficiency: Eugenics, Democracy, and Social Policy in Britain c.1870–1959* (Oxford: Clarendon Press), ch. 4; H. Freeman (2005) 'Psychiatry and the State in Britain' in Gijswijt-Hofstra, Oosterhuis, Vijselaar and Freeman (eds) *Psychiatric Cultures Compared*, pp. 116–140. In addition, the historiography of asylums has highlighted that asylums were never completely segregated from the local communities and society. See for example, R. Porter and D. Wright (eds) (2003) *The Confinement of the Insane: International Perspectives, 1800–1965* (Cambridge: Cambridge University Press); and P. Bartlett and D. Wright (eds) (1999) *Outside the Walls of the Asylum: The History of Care in the Community, 1750–2000* (London: Athlone Press).

[3] H. Oosterhuis (2004) 'Between Institutional Psychiatry and Mental Health Care: Social Psychiatry in the Netherlands, 1916–2000', *Medical History*, 48, 413–28; M. Gijswijt-Hofstra (2005) 'Within and Outside the Walls of the Asylum. Caring for the Dutch Mentally Ill, 1884–2000' in Gijswijt-Hofstra, Oosterhuis, Vijselaar and Freeman (eds) *Psychiatric Cultures Compared*, pp. 35–72; H. Oosterhuis (2005) 'Outpatient Psychiatry and Mental Health Care in the Twentieth Century. International Perspectives', in Gijswijt-Hofstra, Oosterhuis, Vijselaar and Freeman (eds) *Psychiatric Cultures Compared*, pp. 248–274.

[4] N. Henckes (2011) 'Reforming Psychiatric Institutions in the Mid-Twentieth Century: A Framework for Analysis', *History of Psychiatry*, 22, 164–81.

[5] M. Tzanakis (2008) *Beyond the Asylum. Community Psychiatry and the Issue of the Subject* (Athens: Synapsis), pp. 104–107, in Greek. The history of deinstitutionalisation has attracted the interest of few researchers, mostly mental health professionals and to a lesser extent and more recently sociologists and anthropologists. See for example, M. Madianos (1994) *The Psychiatric Reform and its Development. From Theory to Pracrice* (Athens: Ellinika Grammata), in Greek; M. Tzanakis (2003) 'Transformations of Subjectivity and Everyday Life. A Sociological Approach of the Psychiatric Reform in Greece', PhD Thesis, Univeristy of Crete, in Greek; and E. A. Davis (2012) *Bad Souls: Madness and Responsibility in Modern Greece* (Durham, NC: Duke University Press).

practices of inpatient and asylum psychiatry and as local examples of alternative psychiatry and catalysts of the changes that followed in the 1980s.[6] After a brief description of the mental healthcare system in Greece, my analysis turns to these unexplored stories of reform, by examining one of their most active protagonists, the Centre for Mental Health and Research. Based on the writings of the Centre's professionals, as well as on the Centre's cases notes, I analyse the discourses and practices that sought to advance the idea that care had to be relocated from residential institutions to the community. In addition, I explore how the clients and their social environment, primarily their immediate family, responded to and negotiated deinstitutionalisation. My argument is that, while extramural care was hard to implement without state support and the backing of public policy, it was also obstructed by patients and those who made up their social environment, especially parents. Despite these challenges, the Centre's continuous work for 30 years, aided by broader political, social and cultural changes, managed to raise acceptance of community care by the end of the 1970s. Therefore, this case study does not just provide insights into the unmapped history of deinstitutionalisation in Greece; it is also relevant to the larger deinstitutionalisation processes in the post-war world, highlighting the frustrations and achievements of community care.

Deinstitutionalisation in Greece

Deinstitutionalisation in Greece officially began in 1984 with funding from the European Economic Community and was based on the moderate decrease of psychiatric hospital beds, the reform and gradual shutting down of mental hospitals and the foundation and multiplication of alternative services: mental health centres, services for children and adolescents, psychiatric wings in general hospitals, centres for professional training and sheltered homes. Designed on the basis of conclusions reached by a committee of Greek and foreign experts, the new policy needs to be understood within the context of the accession of Greece to the European Economic Community (1979–1981) and the establishment of the National Health System in Greece (1983). With the country on a path of convergence

[6] Tzanakis, *Beyond the Asylum*, pp. 104–107. Tzanakis quotes the view of the psychiatrist Kostas Bairaktaris that attempts of reform before the 1980s were based on the personal initiatives of psychiatrists and not on a general policy of mental health. K. Bairaktaris (1994) *Mental Health and Social Intervention* (Athens: Enallaktikes Ekdosis), p. 107, in Greek.

with the members of the European Community, and with the role of the state in welfare and health expanding, dehumanising mental institutions were not to be tolerated.[7] Furthermore, after the fall of the military junta (1967–1974), a spirit of political and social change prevailed which was evident also in the psychiatric field, where professionals and patients were challenging established practices. Within this context reform of the mental health system appeared necessary by the early 1980s.[8]

Until that time conditions were less favourable to deinstitutionalisation. The state was more orientated to building new hospitals to complement the few that existed before the Second World War: four public and one charity hospital, one university clinic and about ten private clinics. After the War, primarily from the late 1950s to the early 1970s, new residential institutions were founded: the first child psychiatric and neurological hospital, three public psychiatric hospitals, and two 'colonies' for the chronically ill, which were supposed to offer occupational and industrial therapy to inmates and to relieve the overcrowded mental hospitals.[9]

[7] The brutal conditions in Greek mental hospitals were exposed and epitomised by the scandal of the Leros hospital in the late 1970s and 1980s. On this island in the southern Aegean Sea a 'colony of psychopaths' was founded in 1957, in theory to treat patients through work therapy, but in reality to relieve the pressure in the existing mental hospitals by admitting patients who suffered from severe and chronic disorders and had no contact with their families. The term 'colony' was abandoned in 1965, when the institution was named the Psychiatric Hospital of Leros, but conditions remained inhumane throughout the period: the hospital's population, counting in the second part of the 1960s and the 1970s more than 2,000 patients, received minimal treatment and care; many were naked and chained and the staff was mostly untrained. These conditions were first revealed nationally in the second half of the 1970s by newly appointed medical staff on Leros, and were internationally exposed by the *London Observer* in a 1989 article. That year marked the official beginning of deinstitutionalisation in Leros. A. Blue (1999) *The Making of Greek Psychiarty* (Athens: Exadas), pp. 23–32 and 313–14, in Greek; M. Mitrosyli (2015) *Leros Mental Hospital and Reform. Public Policies, Institution, Community* (Athens: Papazisi), in Greek; C. Goutidis and A. Georgiou (2009) 'Presence in Space and Time. The Island of Leros—A Ring in the Chain of History. Leros State Hospital', paper delivered in the 11th PanHellenic Conference of Health Care Services Management, Chania, http://www.leros-hospital.gr/index.php?categoryid=6, last accessed 22/10/2015, in Greek.

[8] Tzanakis, *Beyond the Asylum*, pp. 108–10, 112–12 and 122–24; Tzanakis, 'Transformations of Subjectivity and Everyday Life', pp. 205–206; Davis, *Bad Souls:* 21–50; and Blue, *The Making of Greek Psychiarty*, pp. 119–46.

[9] D. Ploumpidis (1995) *History of Psychiatry in Greece. Institutions, Asylums and Social Context, 1830–1920* (Athens: Exadas), pp. 166–222, in Greek; M. Fafaliou (1995) *Iera Odos 343. Testimonies from Dromokaitio* (Athens: Kedros), pp. 246–48, in Greek. The idea of colonies for the mentally ill was already expressed in the psychiatric law 6077 of 1934, which

The emphasis on more beds was accompanied only to a limited extent by an effort to modernise psychiatric hospitals. Psychologists and social workers were hired for the first time, and new biological, psychodynamic and social practices were introduced, such as leucotomy, ECT and new medications; psychodrama and individual and group psychotherapies; patients' outings in the community, occupational therapy, patients' clubs, day hospitals and outpatient clinics.[10] These structural changes, however, were rather haphazard. They did not follow a general policy to reorganise mental healthcare, despite such intentions being expressed already in the late 1940s, when a committee was formed to propose a new psychiatric law (the existing law dated from 1862).[11] The fact that this bill was never voted on is indicative of the generally weak role of the state in health and welfare in Greece up to the late 1970s and early 1980s, when the foundations of the welfare state were laid. Before that, the deficiencies of the public sector in social protection were principally compensated for by informal networks based on the family and neighbourhood,[12] as well as by a series of private and semi-private organisations, some of which were funded to varying extents by the state.[13] Mental health was no exception: voluntary associations and private institutions provided care, public education and professional training, and also became sources of reform,

pronounced the public asylum of Athens to be a 'colony of Psychopaths', where agricultural and industrial work would be organised, and allowed for the asylums of Thessaloniki and Chania to become colonies. *The Government Gazette* 78, 21/2/1934, in Greek. However, these statements remained theoretical and the first colonies were founded after the Second World War, with the brutal results evidenced in the case of Leros.

[10] Fafaliou, *Iera Odos 343*, pp. 248–49; G. Lyketsos (1998) *The Novel of My Life* (Athens: Gavriilidis), pp. 296–97, in Greek; P. Kazolea-Tavoulari (2002) *The History of Psychology in Greece, 1880–1987* (Athens: Ellinika Grammata), p. 162, in Greek; E. Koutsi (2001) 'Local Society and Institutions of Confinement. The Case of Leros', PhD Thesis, University of the Aegean, pp. 48–76, in Greek.

[11] Lyketsos, *The Novel of My Life*, pp. 291–295.

[12] V. Rapti (2007) 'The Postwar Greek Welfare Model within the Context of Southern European Welfare' in Gro Hagemann (ed.) *Reciprocity and Redistribution: Work and Welfare Reconsidered* (Pisa: Ed. Plus, Pisa University Press), pp. 43–60.

[13] E. Avdela (2013) *'Youths in Danger'. Supervision, Reformation and Justice for Minors after the War* (Athens: Polis), pp. 392–93, in Greek. For the role of the family and other informal networks in providing care, but also for their limitations and intersections with other sources of support, see P. Horden and R. Smith (eds) (2003) *The Locus of Care: Families, Communities, Institutions, and the Provision of Welfare Since Antiquity* (Taylor & Francis e-Library).

instigating extramural services.[14] These reform initiatives had limited scope and scale, precisely because the public sector did not have a significant role in designing and funding them. However, despite their limitations, some of these initiatives had an effect on the treatment of the mentally ill and played a role in advancing deinstitutionalisation before the 1980s, as the case of the Centre for Mental Health and Research demonstrates.

The Centre for Mental Health and Research

The Centre was founded in 1956 by a group of psychiatrists, psychologists and social workers, most of who had studied abroad, mainly in France and the USA, and were inspired by foreign models of dynamic and social psychiatry, such as the English therapeutic clubs, the Association d' hygiène mentale et de lutte contre l' alcoolisme of the 13th district of Paris, and the community mental health centres of the USA.[15] The Centre was initially funded by a royal organisation, the Royal National Foundation, but in 1970 it was placed under state supervision and secured state funding, which increased in the late 1970s.[16] In the following decade the Centre also received European Community grants and expanded its programmes.[17] Previously its services had been rather limited and unevenly distributed in four cities—most were situated in Athens, fewer in Piraeus and Thessaloniki and just one (and only up to 1969) in Patrai. By the end of the 1970s services comprised three child guidance clinics, three psychiatric consultation departments for adults, three day-care units, a school for mentally retarded children, a unit for autistic children and adolescents, a psychotherapeutic unit, a unit for educational programmes and a unit for research.

[14] Lyketsos, *The Novel of My Life*, pp. 289–317 and A. Kaloutsi (1993) *The Greek Society for the Mental Hygiene and Neuropsychiatry of the Child: Thirty Five Years of Interdisciplinary Contribution to the Field of the Mental Health of the Child and the Adolescent* (Athens: Ellinika Grammata), in Greek.

[15] *Mental Health Section, December 1956–December 1963* (1964) (Athens: Royal National Foundation), p. 3 and *passim*; F. Karapanou (1965) *La Consultation Psychiatrique du Centre d' Hygiene Mentale et de Recherches* (Athens: Centre for Mental Health and Research), pp. 10–12; and E. Liberakis (1966) *Group Techniques in a Therapeutic Club* (Athens: Centre for Mental Health and Research).

[16] Proceedings of the Administration Board of the Centre for Mental Health and Research, 36/76, 29/1/1976; 47/76, 9/12/1976; 48/77, 15/1/1977 and 79, 19/9/1979, archive of the Centre for Mental Health and Research, Athens, in Greek.

[17] M. Kaparoudaki and O. Mpotsari (2006) 'The Pioneering Services' in *Greek Centre for Mental Health and Research. 50 Years* (Athens), pp. 12–13, in Greek.

Services that offered diagnosis and treatment provided patients with regular appointments with the psychiatrists, psychologists, social workers, special educators and occupational therapists, along with various methods of treatment: pharmaceutical (antipsychotics, antidepressants and sedatives), psychotherapeutic (counselling, cognitive therapy, behaviour therapy and analytical therapy) and social (professional rehabilitation, occupational therapy, recreational activities).[18] Despite the eclectic character of the therapeutics implemented, the main theoretical tenets of the Centre's professionals originated from dynamic psychiatry: it asserted that mental health and mental illnesses were dynamic conditions, not stable entities, and that there was no clear line between mental health and illness, with the transition from one to the other being gradual and with neurotic elements being present in every personality.[19]

The Centre's services were small-scale. For example, in 1978 the number of patients in all of its day-care units and psychiatric departments totalled 693,[20] while the eight public psychiatric hospitals of Greece had 7,276 patients in 1979.[21] In addition, the Centre's services were not suitable for all patients: while it treated all types of mental disabilities and diseases, in order to participate in the therapeutic programmes most patients had

[18] E. Liberakis (1979) *Notes on the Theory of Psychotherapy* (Thessaloniki: Centre for Mental Health and Research), p. 14, in Greek; Karapanou, *La Consultation Psychiatrique*, pp. 9–10; and E. Kalogeropoulou and P. Matathia (s.d.) *Study on the Cases of Psychotherapy of the Psychiatric Consultation Department for Adults of the Centre for Mental Health and Research* (Athens: Centre for Mental Health and Research), p. 16, in Greek. For the theory and practice of occupational therapy more generally see C. Paterson (2008) 'A Short History of Occupational Therapy in Psychiatry' in J. Creek and L. Lougher (eds) *Occupational Therapy and Mental Health* (Edinburgh: Churchill Livingston Elsevier), pp. 3–16; J. Laws (2011) 'Crackpots and Basket-Cases: A History of Therapeutic Work and Occupation', *History of the Human Sciences*, 24, 65–81; and V. Long (2013) 'Rethinking Post-War Mental Health Care: Industrial Therapy and the Chronic Mental Patient in Britain', *Social History of Medicine*, 26, 738–58.

[19] E. Liberakis (s.d.) 'What is Mental Health?', unpublished speech, folder Liberakis, archive of the Centre for Mental Health and Research, Thessaloniki, in Greek; and 'Proceedings of the Training Meeting of Collaborators and Staff 1/65, 26/1/65', folder Educational Programme for the 'Office of Collaborators' 1958–1965, archive of the Centre for Mental Health and Research, Thessaloniki, in Greek.

[20] 'Statistics Report CMHR 1978', folder Statistic CMHR, archive of the Centre for Mental Health and Research, Thessaloniki, in Greek.

[21] Bairaktaris, *Mental Health and Social Intervention*, p. 108.

relatively mild forms of disturbances or were in an improving condition.[22] Indeed, the Centre did not aim to cover the total mental healthcare needs of the population in its catchment area. It intended mostly to provide model services for prevention and early treatment of the mentally ill, promotion of the population's mental health, training of professionals, education of the public and, not least, reform of the mental health system on the basis of community care.

Discourses and Practices of Deinstitutionalisation

The Centre's professionals saw their work as an agent for the shift from 'custodial' to social psychiatry, and placed their work within the international movement of deinstitutionalisation.[23] They claimed that the Centre represented the 'modern' treatment of the mentally ill, which was not undertaken 'in big hospitals' and did not presuppose the removal of the patients from their family and community.[24] As an alternative to residential care they proposed outpatient treatment which, while offering patients the necessary psychiatric treatment, also mobilised their potential for recovery, as it forced them to adjust to family and community life.[25] In order to facilitate the social integration of patients, psychiatric work was complemented by welfare work for patients with socioeconomic problems (for example, helping them to obtain medication and work)[26] while, most importantly, the Centre established small-scale sheltered workshops and therapeutic clubs.[27] The latter, like the English 'psychiatric social clubs',

[22] E. Liberakis and E. Dimitriou (1980) 'Day Care at the Social Psychiatry Service of Thessaloniki 1971–1980', paper presented at the 9th PanHellenic Conference of Neurologists-Psychiatrists, Athens, folder Liberakis, archive of the Centre for Mental Health and Research, Thessaloniki, in Greek.

[23] 'Proceedings of the Administration Staff, 9/69, 11/3/69', folder Proceedings of Meetings 1/1/69 to 31/12/76, archive of the Centre for Mental Health and Research, Thessaloniki, in Greek; and Karapanou, *La Consultation Psychiatrique*, p. 13.

[24] *Mental Hygiene Section, December 1956–December 1963* (1964) (Athens, Royal National Foundation), p. 4 in Greek; and *Centre of Therapeutic Pedagogy The Stoupathio* (1963) (Athens: Royal National Foundation), p. 6, in Greek.

[25] Karapanou, *La Consultation Psychiatrique*, pp. 5–6.

[26] As a psychiatrist of the Centre noted, psychiatric and welfare work were combined in social psychiatry services of other countries. Liberakis, *Group Techniques in a Therapeutic Club*.

[27] 'Lists of Participation and Programmes', folder Psychiatric Department from 1/10/65 to 31/12/1968, archive of the Centre for Mental Health and Research, Thessaloniki, in

targeted treatment of the 'asocial';[28] patients were helped to take their first steps in social life by creating social relationships in which they would find acceptance, support and sympathy.[29]

From the early 1970s the therapeutic clubs and sheltered workshops were integrated into the new day-care units of the Centre. These transformed the idea of outpatient care to the more intensive and all-round 'day-care' system. This scheme supported patients who were well enough to live at home and who did not benefit from simple outpatient services due to their social isolation and inability to find work.[30] Day patients attended the units several times a week for different types of treatment—occupational therapy, training, group psychotherapy, entertainment, special education and physiotherapy—while they were constantly monitored by the staff, who recorded meticulously their condition and progress.[31] Even entertainment, such as going out for lunch or having coffee in the unit, was intended as social therapy, which helped patients to maintain their 'self-esteem, optimism and a pleasant social appearance which is very different to that of the typical chronic psychotic'.[32] By being part of a group they learned to express their feelings, offer and receive support, adopt responsibilities and take initiatives. As in all services provided by

Greek; 'Proceedings of Staff Meetings in the Station of Thessaloniki 11/66 of 8/3/1966 and 21/66 of 24/5/1966', folder Proceedings of Staff Meetings 1958–1968, archive of the Centre for Mental Health and Research, Thessaloniki, in Greek. The first therapeutic club of the Centre, established in Thessaloniki in 1965, was probably the first in Greece.

[28] Raymond Lerner (1960) 'The Therapeutic Social Club: Social Rehabilitation for Mental Patients', *International Journal of Social Psychiatry*, 6, 101–14, p. 101. For the increasing importance of social rehabilitation in the 1950s and 1960s see V. Long (2011) '"Often There Is a Good Deal to Be Done, but Socially Rather than Medically": The Psychiatric Social Worker as Social Therapist, 1945–1970', *Medical History*, 55, 223–39.

[29] Proceedings of the Administration Board of the Centre for Mental Health and Research, 11/65, 29/9/1965, in Greek; Liberakis, *Group Techniques in a Therapeutic Club*; and case 1, Day Care Unit of Thessaloniki (henceforth DCUT), archive of the Centre for Mental Health and Research, Thessaloniki, in Greek.

[30] Liberakis and Dimitriou, 'Day Care at the Social Psychiatry Service of Thessaloniki'.

[31] Proceedings of the Administration Board of the Association for the Support of Social Psychiatry, 11 3/5/1971, archive of the Centre for Mental Health and Research, Thessaloniki, in Greek; 'Statistics Report CMHR 1978'; and Liberakis and Dimitriou, 'Day Care at the Social Psychiatry Service of Thessaloniki'.

[32] Liberakis and Dimitriou, 'Day Care at the Social Psychiatry Service of Thessaloniki'.

the Centre the final objective was to form responsible, independent and socially adjusted individuals, who would be useful to society.[33]

How successful were these outpatient and day-care services? As already noted, the Centre's services had a relatively small number of patients. In addition, not all patients diagnosed, and therefore recorded, were treated, and some were referred to residential institutions. In what follows, I will use the Centre's case notes to analyse the limitations and achievements of the Centre's deinstitutionalisation practices, exploring how they were received and negotiated by the patients and their environment.[34]

Negotiations, Limitations and Achievements of Community Care

Case notes featured three protagonists: professionals, patients and parents. These were not equally represented. Professionals, the authors of the case notes, decided what was recorded and had the strongest voice. The views of parents, the members of the patients' social environment who came in closest contact with the professionals, featured in case notes more than the patients themselves. Nevertheless, while the patients' views were the least recorded, especially when they were young children, their reactions to the treatment they received were sometimes noted and can offer some idea of how they viewed community care. Therefore, case notes compensate to some extent for the lack of other source material on patients' and lay attitudes towards treatment and the new forms of extramural care.

Between professionals, patients and parents it was the latter who opposed community care most frequently. Up until the late 1960s, parents of young children occasionally asked their children to be placed in residential institutions, including psychiatric hospitals, reformatories, orphanages and boarding technical schools, for economic reasons: they stressed they did not have the necessary resources to raise them.[35] Other parents saw the 'institution', a generic term they used, as the only option for treatment

[33] Case 451, Psychiatric Consultation Department of Thessaloniki (henceforth PCDT), archive of the Centre for Mental Health and Research, Thessaloniki, in Greek; and cases 5, 7 and 13, DCUT.

[34] The case notes archive comes from the Centre's services in Thessaloniki, which, in contrast to the case notes archives of Athens, Piraeus and Patrai, has survived and was made available to me for the purposes of this research.

[35] In the post-war period orphanages admitted children who had lost one of their parents, if the living parent verified that he or she could not support them.

but also the disciplining of offspring of all ages. Mental hospitals were at this time the prevalent locus of care, while residential institutions appealed to parents who could not handle the patient at home or would not tolerate their behaviour outside the home. This was particularly the case for patients who were too noisy or irritable at home,[36] or who behaved in socially unacceptable ways outside the home. Take for instance 15-year-old Anna, who was examined in the medico-pedagogical service of Thessaloniki in 1972. Her mother reported that she was 'mentally retarded', had 'epileptic seizures' and presented 'behaviour disturbances': she wandered around with men and spoke without restraint, using bad language to members of the family. For these reasons the mother requested a certificate so that Anna would be admitted to a 'special institution'.[37]

The psychiatrist who examined Anna in the child guidance clinic ascertained her behavioural problems, mental retardation and epilepsy, and gave the diagnosis 'adolescent problem'. Nevertheless, he did not refer her to an institution, but prescribed medication and advised treatment at the Centre's day-care unit.[38] Anna started treatment willingly, but after a few days was placed by her parents in a private psychiatric clinic. They thought she would 'calm down' in the clinic, while they would have the time to put bars on the house windows, so that she would not 'run away'.[39] After two months, the mother returned to the Centre and explained that they had taken Anna out of the clinic because she was getting worse 'among the mad'. Anna returned to the unit, but not immediately, as her mother did not have the time to escort her and would not be convinced by the social worker to let her go unescorted, probably fearing she would run away again.[40]

This case highlights that extramural care, which was less restrictive and more voluntary than intramural care, was found lacking by parents who used restraining methods at home, such as locking their children in, and who turned to mental health services to get help in controlling their children.[41] These parents could not always be convinced by the

[36] Case 29, '8/may/79', DCUT.

[37] Case 1, 'Social History', Child Guidance Clinic of Thessaloniki (henceforth CGCT), archive of the Centre for Mental Health and Research, Thessaloniki, in Greek.

[38] Case 1, 'Psychiatric Examination', CGCT.

[39] Case 1, letter to the mother, 27/4/1972 and '11/4/72', CGCT.

[40] Case 1, 'Interview with the mother, 19/6/72', CGCT.

[41] See also case 23, 'Interview with the father for the announcement of results' and 'Interview with the mother, 22/2/78', DCUT.

Centre's professionals on the suitability of outpatient care, as Anna's case indicates. However, the professionals generally tried to deter parents from residential care and, drawing on psychiatric and sociological studies on the effects of institutional life, they rarely advised admission to a residential institution.[42] This was seen only as the last resort when community care methods had failed and the patients' condition had become critical and beyond improvement with outpatient services. For example, 15-year-old Yannis, who was diagnosed with schizophrenia, was treated for three years in the day-care unit in Thessaloniki,[43] before the psychiatrist recommended that he should be admitted to a mental hospital. He had deteriorated, manifesting 'psychotic stress', decline of mental abilities, emotional disturbances and many irrational ideas. Under these circumstances the staff could not reach him on any level and turned their efforts to suppressing his symptoms 'with the hope of avoiding inpatient hospital care'. This became impossible though, as he became ill with an infectious disease which complicated his condition.[44]

In addition, hospitalisation seemed inevitable when there was no family or when the family was judged as harmful or as unable to support community care. For instance, 24-year-old Alex, diagnosed with 'unspecified psychosis', was treated in the unit of Thessaloniki from 1976 to 1981. During this period the patient was staying with his father, who saw the social worker regularly and participated in a parents' group. In 1981, however, the professionals advised Alex's admission to a hospital, as he was 'on his

[42] See for example A. Freud and D. Burlingham (1973) *Infants Without Families. Reports on the Hampstead Nurseries, 1939–1945, The Writings of Anna Freud,* v. 3 (New York: International Universities Press); R. Spitz (1945) 'Hospitalism. An Inquiry into the Genesis of Psychiatric Conditions in Early Childhood', *Psychoanalytic Study of the Child*, 1, 53–74; J. Bowlby (1950) *Maternal Care and Mental Health* (Geneva, World Health Organization, monograph series no. 3); E. Goffman (1961) *Asylums; Essays on the Social Situation of Mental Patients and Other Inmates* (Garden City, NY: Anchor Books). The psychiatrist of the Psychiatric Department of Thessaloniki, Eustathios Liberakis, had analysed in his PhD thesis the issue of institutionalisation. E. A. Liberakis (1978) 'Some Factors Predisposing to Institutionalism in Chronic Psychiatric Patients', PhD Thesis, Memorial University of Newfoundland, http://thesis.ekt.gr/thesisBookReader/id/3781#page/1/mode/2up, last accessed 25/5/2015.

[43] Case 23, 'Interview with the father for the announcement of results' and 'Interview with the mother, 22/2/78', DCUT.

[44] Case 23, 'Information Note, 31/1/1979', DCUT.

own, without money, without supervision'. He had not been taking his medication, had relapsed and could not be helped in the unit.[45]

That the family had always played a significant role in psychiatric treatment has been a well-known fact, even in the case of residential care.[46] What Alex's case demonstrates is that community care also presupposed that those who made up the social environment of the patients, especially parents, participated in treatment. Participation included supervising patients, ensuring, for example, that they took their medication, but also being actively involved in therapy: parents were expected to cooperate with professionals, follow their advice and transfer the Centre's therapeutic principles to the home context. For instance, parents were expected to understand patients' problems in behaviour as stemming from emotional troubles within the family and to encourage patients to talk freely about their feelings, needs and desires. In addition, parents were advised not to try to control the patients with strict prohibitions, but to set boundaries and encourage them to take initiatives and become responsible and independent.[47] In these ways the therapeutic aims of the Centre's services (responsibility, independence and social integration) were to be equally supported by the patients' families. By making these demands on the environment of the patients, outpatient care was limited to patients with family or friends willing and able to undertake roles of supervision and therapy.

[45] Case 25, letter of the psychiatrist to the Alex's sister, 9/7/1981, DCUT.

[46] M. Finnane (1985) 'Asylums, Families and the State', *History Workshop Journal*, 20, 134–48; N. Tomes (1994) *The Art of Asylum-Keeping: Thomas Story Kirkbride and the Origins of American Psychiatry* (Philadelphia: University of Pennsylvania Press), pp. 90–124; J. Walton (1985) 'Casting out and Bringing back in Victorian England: Pauper Lunatics' in W. F. Bynum, R. Porter and M. Shepherd (eds) *The Anatomy of Madness: Essays in the History of Psychiatry*, v. 2 (London: Tavistock), pp. 132–46; C. McGovern (1987) 'The Community, the Hospital, and the Working-Class Patient: The Multiple Uses of Asylum in Nineteenth Century America', *Pennsylvania History*, 54, 17–33; P. Prestwich (1994) 'Family Strategies and Medical Power: "Voluntary" Committal in a Parisian Asylum, 1876–1914', *Journal of Social History*, 27, 799–814; D. Wright (1997) 'Getting out of the Asylum: Understanding the Confinement of the Insane in the Nineteenth Century', *Social History of Medicine*, 10, 137–55; and L. Wannell (2007) 'Patients' Relatives and Psychiatric Doctors: Letter Writing in the York Retreat, 1875–1910', *Social History of Medicine*, 20, 297–313. For the importance of the family in the care and control of the mentally ill and disabled see A. Suzuki (2003) 'The Household and the Care of Lunatics in Eighteenth-Century London' in Horden and Smith (eds) *The Locus of Care*, pp. 153–75; and D. Wright (2003) 'Familial Care of "Idiot" Children in Victorian England' in Horden and Smith (eds) *The Locus of Care*, pp. 176–97.

[47] Case 1181, 'Interview with the youth herself 24/11/75', PCDT.

Parents often could not or would not meet the challenge of 'soft' supervision and resisted the liberal and psychological model of family relationships proposed by the Centre. Thus they could not make community care work and sometimes turned instead to residential institutions.

Patients could also reject the theoretical and practical terms of outpatient treatment. During the 1960s they often resisted psychotherapies and insisted on getting 'medical' treatment, such as pills and shots. In addition, they did not acknowledge the importance of attending appointments regularly and did not always follow the professionals' instructions. These reactions definitely posed obstacles to outpatient treatment, which was based to a great extent on the willing and active participation of patients. In the 1960s some professionals attributed poor patient cooperation to their low socioeconomic background.[48] However, even in the next decade, when the number of more well-off patients had increased,[49] there were still many patients who did not get along with the modus operandi of the services and either stopped attending on their own initiative or were let go by the professionals.

Despite these negotiations of and objections to community care, there were also parents and patients who were willing to be subjected to the regime and adopt its principles. Indeed, in the course of time patients' and parents' 'resistance' to the therapeutic model of the Centre diminished. As the Centre's case notes indicate, by the end of the 1970s parents were more easily deterred from institutional solutions, while patients seemed more favourable to the interpretations and dynamic and social methods of the services. For example, during the first meeting of the psychotherapy group in 1972 a patient said that she had decided to participate in the group to learn more about herself and become a more mature woman.[50] Others patients announced that they wished to transfer day-care methods, such as the structured organisation of their daily programme, to their life outside the services.[51]

This change in the attitudes of patients and parents can be attributed only partially to the Centre's efforts to familiarise them with the prac-

[48] Liberakis, *Group Techniques in a Therapeutic Club*.

[49] This is indicated, for example, by the increased number of patients who could afford to pay for their treatment in the 1970s.

[50] 'Proceedings of group psychotherapy, 7/7/1972', folder Group Psychotherapy of Women, Therapeutic Club, Parents' Group, DCU, Researches, from 1/1/1972 to 31/12/1972, archive of the Centre for Mental Health and Research, Thessaloniki, in Greek.

[51] Case 12, interview with the social worker, 31/5/71, DCUT.

tical and theoretical requirements of outpatient treatment and to train them to cooperate in the various therapies. To a greater extent it was influenced by broader social changes that had been underway since the 1950s and 1960s and which, by the 1970s, were having a cumulative impact: urban populations were increasing and living standards improving; European and American social movements were more commonly known, and democratisation and liberalisation demands, especially from the youth, were intensifying. Although the wave of social activism of the 1960s was interrupted between 1967 and 1974, when Greece was controlled by a military dictatorship, the following period was highly politicised, with a plethora of attempts to reform the political system and social institutions.[52]

These changes liberalised social and family relationships, which made parents more favourable towards less restraining treatments and, thus, outpatient care. In addition, as mental health services and professionals increased, at least in urban Greece, psychological discourses like those of the Centre, proposing a liberal and psychological family model became more widespread. Within this social, political and cultural framework, it is not strange that by the end of the 1970s fewer parents turned to the Centre's services to control their children, and that professionals had more chances to convince parents of the unsuitability of intramural care and to promote community care.

With regard to patients, the extent to which social, political and cultural conditions were combined with the work of the Centre to influence their attitudes is illustrated by the formation of a dynamic group of patients within the day-care unit of Athens in 1979. In October a number of patients protested against what they called 'arbitrary interventions' of the Centre's administration in the relationships between therapists and patients. They referred to the decision of the administration to shut down the psychotherapeutic unit and to move a psychiatrist from the day-care

[52] For the social changes in Greece between the 1950s and 1970s see Avdela, *'Youths in Danger'*; K. Kornetis (2013) *Children of the Dictatorship: Student Resistance, Cultural Politics and the Long 1960s in Greece* (New York: Berghahn Books); K. Katsapis (2013) *The 'Problem Youth'. Modern Youths, Tradition and Dispute in Post-War Greece: 1964–1974* (Athens: Aprovleptes Ekdosis), in Greek; and N. Papadogiannis (2010) 'Greek Communist Youth and the Politicisation of Leisure, 1974–1981', PhD Thesis, University of Cambridge. Certainly, liberalisation was not a straightforward and continuous process, but it can be argued that in the 1960s and the 1970s liberal ideas and practices were gaining ground against more traditional and autarchic ones.

unit of Athens to that of Piraeus.[53] Although their reaction did not halt these measures, their experience of collective action led them to form the first mental patients association in Greece, the Movement for the Rights of the 'Mentally Ill', in February 1980. The analogy with the way in which the first mental patients' association in England was founded is striking: the creation of Mental Patients Union in 1972 originated from a group of patients and professionals who had opposed the impending closure of the Paddington Day Hospital in London.[54]

The major aims of the Movement for the Rights of the 'Mentally Ill' were to inform the public on mental health issues, defend the rights of the mentally ill and improve the mental healthcare system in Greece. It was a radical organisation which drew its inspiration from antipsychiatry, as manifested by the quotation marks around the term 'Mentally Ill' in the name of the association. Its magazine, *Freedom is Therapeutic*, was equally radical, containing translations of texts by R. D. Laing and David Cooper. The Movement stressed the evils of mental health services in Greece and argued that psychiatric power and every form of confinement had to be abolished.[55]

The radical character of the association was certainly shaped by the more liberal climate of the late 1970s; it is indicative that during this time the ideas of antipsychiatry started to circulate more widely in Greece.[56] However, along with being part of wider political, social and cultural changes, the Movement originated as a patients' initiative within a day-care

[53] Proceedings of the Administration Board of the Centre for Mental Health and Research, 79, 19/9/1979; 80, 17/10/1979; and 81, 7/11/1979.

[54] N. Crossley (2006) *Contesting Psychiatry: Social Movements in Mental Health* (London: Routledge), pp. 128–44. In Greece, the Movement for the Rights of the 'Mentally Ill' was also created by both patients and professionals, as were patients' organisations that were founded in the 1970s in other countries, such as Canada, Germany and the Netherlands. Gijswijt-Hofstra, 'Within and Outside the Walls of the Asylum'; V. Roelcke (2005) 'Continuities or Ruptures? Concepts, Institutions and Contexts of Twentieth-Century German Psychiatry and Mental Health Care' in Gijswijt-Hofstra, Oosterhuis, Vijselaar and Freeman (eds) *Psychiatric Cultures Compared*, pp. 162–82; and M. Davies et al., 'After the Asylum in Canada: Surviving Deinstitutionalisation and Revising History', ch. 4 in this volume.

[55] 'The Handling of the "Mentally Ill": Institutions and Mentalities' (April 1983), *Freedom is Therapeutic*, 2, 3–8 (in Greek).

[56] Although antipsychiatry in Greece has not been studied systematically, the movement was becoming more prominent in the latter half of the 1970s, when the first antipsychiatric magazine, *Madness*, was published. Tzanakis, 'Transformations of Subjectivity and Everyday Life', pp. 205–206.

unit of the Centre. The way that these services worked—without confinement, based on voluntary treatment and group participation—was conducive to a new kind of relationship between patients and professionals; to a new kind of balance between them, which made possible the articulation of patients' demands.

Conclusions

To conclude, the Centre for Mental Health and Research is a valuable example of the process of deinstitutionalisation. Beyond its national context, it illustrates that private initiatives could play a part in this process; that the clinical, familial and social conditions of a patient's life complicated the implementation of community care and that the borders between inpatient and outpatient care were porous, with patients going back and forth between outpatient services and hospitals. On the national level, this particular case study demonstrates the limited practical impact of deinstitutionalisation in Greece between the 1950s and 1970s. Even the Centre, one of the first, longest-lived and most active agents of community care, faced various challenges, having limited funding and public support up to the late 1970s and encountering opposition from patients and their environment for various—economic, social and cultural—reasons. Within this context, the Centre remained an isolated agent of community care until the 1980s. It covered just a small minority of the overall number of those in mental healthcare and usually the milder cases; it did not manage to establish community care as an equal alternative to residential care; and, even more, it did not trigger, as it aspired to, reform of the mental healthcare system in Greece.

Even so, the Centre, based on foreign deinstitutionalisation models, did offer an alternative form of care on the basis of voluntary outpatient treatment, the participation of the patients and their social environment, and the combination of medical, psychological and social methods. Since patients stayed at home to avoid the isolation and marginalisation associated with residential care, both they and the people around them were required to cooperate willingly and actively with the professionals, undertake responsibilities regarding treatment and, more generally, accept the therapeutic model that the Centre promoted. This model was not based on external disciplining, restriction and isolation, but on self-control, independence, responsibility and integration, and presupposed a liberal and psychological model of family and social relationships.

By the end of the 1970s patients and parents were becoming more positive towards this therapeutic and social model, while patients were starting to question hierarchies that had been long established within psychiatric institutions. Certainly this change was made possible by the more liberal climate and the greater dispersion of psychological methods during the 1970s. At the same time, however, it was activated by the impact exercised on patients and their families by the Centre's community work for 30 years. By familiarising the patients and their families with outpatient treatment and its liberal and psychological principles, the Centre succeeded in shaping and disseminating a new mentality of mental healthcare—a mentality that was compatible with and promoted community care several years before the beginning of deinstitutionalisation in Greece. It was by moulding and activating this new mentality that even an isolated agent of deinstitutionalisation such as the Centre for Mental Health and Research contributed to the psychiatric reform of the following years.

PART III

New Conceptualisations of Therapy and Space

CHAPTER 9

Child Guidance and Deinstitutionalisation in Post-War Britain

John Stewart

Introduction

This chapter examines the implications of child guidance, an influential movement in child mental welfare, for deinstitutionalisation in Britain. While it is not suggested that child guidance provided the only intellectual critique of institutional care it did, nonetheless, help create a climate wherein the whole notion of institutionalisation came under scrutiny. This was because child guidance proposed the centrality of the family to child rearing, however apparently flawed some families might be. Supporting the family was to be done by way of child guidance clinics which were posited as an alternative to what were seen as old-fashioned, impersonal and uncaring, and thus psychologically damaging, institutions such as children's homes. Although cross-national comparisons are challenging, similar approaches appear to have had similar outcomes in other countries. Kari Ludvigsen and Åsmund Arup Seip argue that in post-war Norway child guidance was central to a child welfare policy promoting

J. Stewart (✉)
Glasgow Caledonian University, Glasgow, UK
e-mail: John.Stewart@gcu.ac.uk

D. Kritsotaki et al. (eds.), *Deinstitutionalisation and After*,
Mental Health in Historical Perspective,
DOI 10.1007/978-3-319-45360-6_9

'integration' at the increasing expense of 'segregation'.[1] In Ireland in the same period child guidance provided a different model for the treatment of maladjusted children in a society where previously the predominant approach had involved institutionalisation and moral censure rather than the family-based and non-judgemental support which child guidance claimed to advance.[2]

Child guidance's historical origins are also important here. When founded in the interwar period it was part of the international movement for mental hygiene and the desire of psychiatrists, in a number of countries, to reconstitute their professional aims and practice. Discussing the United States, in a passage revealingly entitled 'Prelude to de-institutionalisation', Gerald Grob remarks that by identifying with mainstream medical science 'psychiatrists slowly began to shift the location of their practice from mental hospitals to outpatient facilities, child guidance clinics, and private practice'.[3] Such aspirations were held from the outset by British child guidance practitioners, many of the first generation of whom had first-hand experience of American clinics, and received further impetus by way of social reconstruction in the latter half of the 1940s.

Chronologically we begin in the aftermath of the Second World War when one component of the emerging welfare state was the 1948 Children Act. This stressed the desirability of locating the care of children wherever possible in the family home, and in any event not in large-scale institutions. As Harry Hendrick points out, one of the key factors shaping the Act was the child guidance movement and its well-placed supporters. The Act derived from the findings of the recent Care of Children (Curtis) Committee whose members included leading child guidance practitioners, psychiatric social worker Sybil Clement Brown and psychologist Lucy Fildes. Those giving evidence included individuals with strong and long-standing child guidance connections, notably psychiatrist John Bowlby,

[1] K. Ludvigsen and Å. A. Seip (2009) 'The Establishing of Norwegian Child Psychiatry: Ideas, Pioneers, and Institutions', *History of Psychiatry*, 20:1, 5–26.

[2] T. Feeney (2012) 'Church, State, and Family: The Advent of Child Guidance Clinics in Independent Ireland', *Social History of Medicine*, 25:4, 848–62.

[3] G. Grob (2005) 'The Transformation of Mental Health Policy in Twentieth-Century America' in M. Gijswijt-Hofstra, H. Oosterhuis, J. Vijselaar and H. Freeman (eds) *Psychiatric Cultures Compared: Psychiatry and Mental Health Care in the Twentieth Century* (Amsterdam: Amsterdam University Press), pp. 141–61, 144.

part of whose training had been at the London Child Guidance Clinic; psychiatric social worker Robina Addis; and social reformer Evelyn Fox.[4]

This chapter's end point is the early 1970s after which there were important changes in social work practice and philosophy and, soon after, in child and family policy. In part these changes arose from the 'rediscovery of poverty' in the 1960s. By this point too the validity of the child guidance approach was beginning to be called into question, part of what Mathew Thomson calls a 'collapse of confidence in home and family'.[5] Again the broader context illuminates the particular issue of child mental welfare. As John Welshman remarks, the Second World War had served 'to focus attention on the mental health needs of both children and adults'. In principle, if not necessarily immediately in practice, the advent of the National Health Service (NHS) in 1948 sought to address such needs and with respect to the following 25 years Welshman identifies the major landmarks on the road to community-based, rather than institutional, mental healthcare for all age groups, including the Seebohm Report, cited below.[6]

What then was child guidance and what did it seek to achieve? Child guidance was a form of psychiatric, preventive medicine aimed at children from birth to adolescence. It originated in the USA, came to Britain in the late 1920s, and expanded its influence and practice to such an extent that it was legislatively embedded in the post-war welfare state. The underlying premise was that any child, however apparently 'normal', could become, in the course of their emotional and psychological development, maladjusted and that this had consequences not just for the child but also for her or his family and society as a whole. Maladjustment manifested itself through a variety of behavioural disorders, ranging from bed-wetting to shyness. The locus for diagnosis was the child guidance clinic where the child would encounter members of three professions—psychiatry, psychology and psychiatric social

[4] H. Hendrick (2003) *Child Welfare: Historical Dimensions, Contemporary Debate* (Bristol: The Policy Press), ch. 4; The Home Department and the Ministry of Education (1946) *Report of the Care of Children Committee, Cmd.6922* (London: HMSO). On child guidance, J. Stewart (2013) *Child Guidance in Britain, 1918–1955: The Dangerous Age of Childhood* (London: Pickering and Chatto).

[5] M. Thomson (2013) *Lost Freedom: The Landscape of the Child and the British Post-War Settlement* (Oxford: Oxford University Press), p. 101.

[6] J. Welshman (1999) 'Rhetoric and Reality: Community Care in England and Wales, 1948–74' in P. Bartlett and D. Wright (eds) *Outside the Walls of the Asylum: The History of Care in the Community, 1750–2000* (London: The Athlone Press), pp. 204–26, 207–8, 213–14.

work. Most treatment took place in the home, effectively putting the psychiatric social worker on the front line. The purported causes of maladjustment are discussed below. Child guidance was important in pathologising childhood which was both inherently problematic but also, because of the nature of the child psyche, a time when intervention in a non-institutional setting by appropriately trained experts could yield positive results. Maladjusted children were thus sick, not morally tainted. Child guidance also stressed that each child was unique—biologically, psychologically and emotionally—and had to be seen in his or her total context. We now take extracts from publications drawn from across the quarter century after 1945. These are used to draw out key issues which are then further elaborated upon.

Outlining the Issues

Our first extracts come from the psychologist Agatha Bowley, in the mid-1940s a child guidance practitioner and thereafter a lecturer in childcare at the University of London. In 1946 she argued that there were three underlying principles to happy family life: a good relationship between husband and wife; a good relationship between parents, and especially the mother and children; and a good relationship between siblings. Regarding the child and its parents, the 'careful wife' would not allow the child to come between her and her husband and it had to be recognised that jealousy was all too possible—the husband might be jealous of the child or the child of its father. On siblings, like many of her contemporaries Bowley thought that a one-child family was psychologically problematic. 'The large family, or the family of two or three', she argued, had an 'advantage over the only child, where feelings are often intense and highly concentrated'. Specifically for children, there were five fundamental psychological conditions which had to be fulfilled if 'successful adjustment to the environment is to be achieved'. These were affection; security, by which she meant emotional security; freedom; discipline; and emotional, social and intellectual outlets.[7]

A few years later Bowley further elaborated her ideas. The family's 'special function' was to 'provide security, affection and adequate care for the child, until he is able to fend for himself [*sic*]'. Children without one or

[7] A. Bowley (1946) *The Problems of Family Life: An Environmental Study* (Edinburgh: E & S Livingstone), pp. 2–5, 7, 20ff.

both parents faced psychological problems. Most at risk were children under four who lost their mothers, this having 'the most serious effect on emotional development'. Periods of separation from the mother or the child's prolonged hospitalisation could produce similar results. Although she clearly thought that such measures were psychologically second-best, Bowley acknowledged that some children would have to be fostered or placed in institutions. As things presently stood, though, institutions were too large, unnecessarily segregated by sex and age, overemphasised cleanliness at the expense of happiness, and lacked good personal relationships between children and staff.[8] Cleanliness came up frequently in this sort of discourse, the argument being that an obsession with it symbolised a sort of emotional sterility which contrasted with messy, but emotionally healthy, normal family life.

The second set of extracts comes from a National Association for Mental Health (NAMH) publication of the mid-1960s. This voluntary body actively promoted child guidance—one of its predecessor organisations was the Child Guidance Council—as well as publicly campaigning for better care for the mentally ill of all ages and for community- rather than hospital-based provision. Indeed it had itself pioneered certain community-based projects. The organisation also had a network of local associations and strong international links.[9] The pamphlet's format, obviously aimed at the general public and clearly popular as it ran to several editions, was a question and answer structure. So, for example, to the question 'Is mental illness inherited?' it was conceded that mental illness did run in families but most psychiatrists were 'now inclined to attribute mental illness to environmental factors or personal experiences'. As to prevention, 'probably the biggest single factor in ensuring mental health' was emotional security in childhood, hence the need for parents to 'provide a stable, loving environment for their children'. A third question concerned 'broken homes' and whether they 'inevitably' led to mental instability. The response was that while many survived 'very unfortunate childhood experiences without becoming mentally ill' there was nonetheless 'little doubt that an insecure family background increases the chance of mental breakdown'.[10]

[8] A. Bowley (1949) *Psychological Aspects of Child Care* (Liverpool: Liverpool University Press), pp. 3, 6, 13, 19.

[9] N. Crossley (1998) 'Transforming the Mental Health Field: The Early History of the National Association for Mental Health', *Sociology of Health and Illness*, 20:4, 458–88.

[10] National Association for Mental Health (1966) *Questions on Our Minds*, 3rd edn (London: NAMH), pp. 3, 5, 7, 11.

The third set of quotes comes from the tenth edition of a standard, if increasingly dated, psychiatric text published in 1969 and usually known by the surnames of its original authors, Henderson, David Kennedy and Gillespie, Robert Dick, both advocates of child guidance before 1939. One of the key issues in child psychology, it claimed, was the 'enormous influence of environment on the child's mental processes', that is 'the personal environment of parents, brothers, sisters, teachers and companions'. Because of the nature of the child's mind 'morbidity' was more usually attributable to 'some direct frustration' rather than 'intrapsychic conflict'. Consequently 'relatively simple adjustments of the environment' would generally produce a cure and this could often be done by sessions with the parents rather than the child. Regarding parents, it was a 'commonplace that the relationship of parent to child is of maximal importance for the shaping of character'. But equally important was that relationship's role in the 'production of psychopathological deviations in childhood'. Common problems included 'the domineering father over-keen for the progress and advancement of his son'. Parents' unhappiness with their own lives too could be problematic. So, for instance, a girl who felt unloved by her mother might overcompensate when she had children of her own.[11] It was central to the child guidance approach that bad practice could be passed down the generations. Poor parenting thus had deleterious, and potentially long-term, consequences.

The Problem and Its Origins

What can we make of these extracts? In each case, and typically, stress was laid on the child's environment. But what environment did not usually mean was the socioeconomic circumstances in which children found themselves. Rather, environment was the emotional landscape of the child's family, and particularly the quality of the parent/child relationship. Good parenting led to stable, adjusted children who would contribute positively to society. Bad parenting, though, led to maladjustment and longer-term individual and social problems.

Of course material circumstances could not be entirely ignored. Henderson and Gillespie conceded that matters such as 'economic status' were often of 'much importance' in childhood maladjustment, although

[11] *Henderson and Gillespie's Textbook of Psychiatry* (revised by I. R.C. Batchelor) (1969) 10th edn (London: Oxford University Press), pp. 473, 479–80.

that was the extent of the volume's engagement with the issue.[12] Bowlby claimed in 1953 that real poverty no longer existed in Britain thanks to full employment and the NHS. Like many others at this time, he believed that poverty had been more or less abolished by the welfare state. But poverty could play its part in psychological damage to children as in the case of a mother with no husband and thereby unable both to care for her young children and earn a decent living. In such circumstances the children might be taken into care. Bowlby, though, argued that the remedy lay in spending on social services so that the family might remain together. To do otherwise was, other considerations aside, bad economics. Similarly, it was 'axiomatic' that children should not be removed from their home on account of the mother's illness for 'it is less expensive to put in a home help'. Similarly, children should not be removed from home 'simply because it is dirty'. It was cheaper to 'put in a family service unit'.[13] Bowlby, in acknowledging a particular case of deprivation, was simultaneously emphasising the importance of the mother/child relationship and arguing for social service provision to avoid the need for institutional or foster care. Bowlby was hugely influential in post-war British child-rearing and is a recurring figure in our narrative.[14]

It is important to emphasise that child emotional or psychological maladjustment was not seen as an insignificant or a purely individual problem. The Ministry of Education's Committee on Maladjusted Children (the Underwood Committee) claimed in its 1955 report that at any one time around ten per cent of the child population had, or might develop, some sort of problem. On socioeconomic circumstances, the Committee remarked that although these might have some impact, nonetheless 'many fine men and women have been reared in slum dwellings or in rural homes at bare subsistence level'. If children in such conditions received 'the right kind' of parental care and affection no maladjustment would occur. By contrast, there was 'much evidence that failure in personal relationships is the most important factor in maladjustment'.[15] An NAMH conference dedicated to the Underwood Report attracted over 800 people, the largest such event yet staged by the organisation. It also came in the wake of

[12] Ibid., p. 483.

[13] J. Bowlby (1953) *The Roots of Parenthood* (London: National Children's Home), pp. 11–13.

[14] For a recent discussion, Thomson, *Lost Freedom*, ch. 3.

[15] Ministry of Education (1955) *Report of the Committee on Maladjusted Children* (London: HMSO), ch. 13, 'The Size of the Problem', pp. 29, 28.

the BBC television series 'The Hurt Mind', one strand of which covered child guidance and in which the NAMH had been actively involved.[16] The Underwood Report's impact may also explain the twofold increase between 1954 and 1958 in those attending the annual conference for child guidance professionals.[17]

In the late 1960s the government enquiry into social work, generally known as the Seebohm Committee, noted that around 55,000 children were seen annually at child guidance clinics. Such services were, though, still inadequate with research suggesting that at least one child in ten would need special educational, psychiatric, or social help before the age of 18 while at most one child in twenty-two was presently receiving appropriate care. Adequate levels of child care were an 'investment in the future' and it made no sense either on 'humanitarian grounds or in terms of sheer economics' to allow children's physical, intellectual, or emotional neglect. To do so would 'mortgage the happiness of thousands of children' and the offspring they in turn would have as well as piling up 'future expense and problems for society into the bargain'.[18]

Emphasis on the potential impact of child maladjustment was part of a broader post-war belief, reinforced by such psychological analyses that social and political stability depended on familial stability—hence the argument that wherever possible children should remain with their families rather than being placed in institutions. Proponents of such beliefs further argued that interventions like child guidance were necessary to counter instances of excessive parental licence or excessive parental control—both potentially problematic and thereby both potentially disruptive. There was a 'middle way' in child-rearing just as in the purported post-war consensus over social and economic affairs which sought to achieve stability in the aftermath of a devastating war and the upheavals of the interwar era. Advocates of child guidance had seen it as central to wartime plans for

[16] Wellcome Library, Archives and Special Collections, Mind Archive (hereafter, Wellcome), SAMIN/A/2/2, Minutes of NAMH Council, 26 April 1957, 'Appendix A, Report from the General Secretary', p. 1. On 'The Hurt Mind' see V. Long (2014) *Destigmatising Mental Illness? Professional Politics and Public Education in Britain, 1870–1970* (Manchester: Manchester University Press), p. 202ff.

[17] Wellcome, SAMIN/A/2/2, Minutes of NAMH Council, 18 April 1958, 'Report from General Secretary', p. 4.

[18] Secretary of State for the Home Department, Secretary of State for Education and Science, Minister of Housing and Local Government, Minister of Health (1968) *Report of the Committee on Local Authority and Allied Personal Services* (London: HMSO), pp. 53, 58.

post-war social reconstruction and its implementation and practice as a mild form of social democracy, an issue returned to below.[19]

Supporting Children and Parents

If children's behavioural problems were attributable to a dysfunctional environment and had potentially detrimental long-term consequences, what was to be done? In the past, the default response would have been to take children out of such an environment and put them into some form of institutional care. But such an approach was not only questionable in itself; it also capitulated in the face of the problem. The child guidance clinic psychiatrist Kate Friedlander claimed in the late 1940s that there was now enough scientific evidence to confirm that 'the family unit is the best medium for a satisfactory emotional and character development' and that this had informed the Curtis Committee's work.[20] A standard social administration text suggested in the mid-1950s that until recently the maladjusted child had been viewed as requiring punishment but was now seen as 'sick' and 'in need of treatment'. So while residential facilities might be required for the more seriously troubled it was the 'child guidance clinic which was the most potent factor in dealing with the problem of maladjustment'.[21] A few years earlier the psychiatrist and leading child guidance campaigner Douglas MacCalman told an international meeting that at clinics even the worst parents responded to 'careful influence' and there was now much evidence that the 'bad parent was better than none at all'.[22]

This echoed the Curtis Committee's comments that at least some basic childhood needs were supplied by the family home even if 'in some respects it is not good one'. It was thus a 'very serious responsibility' to ensure that if taken out of the family then such basic needs could be supplied by 'the environment to which he is removed'.[23] And even if relocation to a special

[19] Stewart, *Child Guidance*, chs 6, 9. Also, Thomson, *Lost Freedom*, ch. 3.

[20] K. Friedlander (1948) 'The Significance of the Home for the Child's Emotional Development during the First Six Years', *Journal of Mental Science*, 94, April, 305.

[21] M. P. Hall (1955) *The Social Services of Modern England*, 3rd edn (London: Routledge and Kegan Paul), pp. 185, 188.

[22] 'International Congress on Mental Health' (1948) *British Medical Journal*, 2:4573, 396.

[23] The Home Department and the Ministry of Education, *Report of the Care of Children Committee*, p. 139.

school was absolutely necessary it was with one outcome in mind. As the child guidance psychiatrist Dr W. Whiles told a conference in 1958 the 'aim of any residential placement' was 'the return of the child to his family'. Residential schools were not substitute homes but rather somewhere the child could learn to see his or her parents 'more objectively' with a view to once again being 'part of the family'. The child guidance clinic, meanwhile, should see the child during school holidays and the parents throughout the year. Whiles cited the Underwood Report's stress on the family as an 'anchorage' to which the child should ultimately return.[24]

The 'evidence' that the 'bad parent was better than none at all' was being gathered by Bowlby, one result being his report of the early 1950s for the World Health Organization. This argued that during the preceding 25 years one of the 'most significant developments in psychiatry' had been the 'steady growth of evidence that the quality of the parental care which a child receives in his earliest years is of vital importance for his future mental health'. This had come first from psychoanalysts and then from psychiatrists and psychologists working in childcare and child guidance. Bowlby emphasised that separation from the mother need not necessarily be physical were she, for any reason, 'unable to give him [*sic*] the loving care small children need'. But even staying in a loveless home or being fostered out was better than 'the almost complete deprivation' still to be found 'in institutions, residential nurseries, and hospitals, where the child often has no one person who cares for him [*sic*] in a personal way and with whom he may feel secure'. This was why 'children thrive better in bad homes than in good institutions and why children with bad parents are, apparently unreasonably, so attached to them'. Among those who had reached this conclusion were 'most experienced social workers with mental health training'.[25] Bowlby's hostility to institutions is striking, was widely shared among child guidance practitioners and supporters, and significant given his rising professional and public status.

So parents, more accurately mothers, were much to be preferred for child-rearing but might have to be told how. Bowlby, like others associated with child guidance, sought to spread his message beyond academic

[24] Wellcome, SAMIN/B/64, Dr W. H. Whiles (1948) 'Co-Operation between Clinic and Schools for the Maladjusted' in National Association for Mental Health, *The Residential Care of Disturbed Children: Being the Proceedings of the 14th Inter-Clinic Conference* (London: NAMH), pp. 20–21, 17, 12–13.

[25] J. Bowlby (1952) *Maternal Care and Mental Health*, 2nd edn (Geneva: World Health Organization), pp. 11–12, 68.

and practitioner audiences. His most influential work, *Child Care and the Growth of Love*, explained 'attachment theory' for a general readership. Attachment theory argued that the early years were crucial to child development, particularly through the creation of a bond between mother and child. Bowlby also suggested that child guidance workers acknowledged that they now 'give as much time to the treatment of the parents as to that of the children'. European and American opinion, moreover, now held that removing the child from the home was unadvisable and so the 'provision of child guidance services on a generous scale' was the first step towards 'the maintenance of family life' and, thereby, 'the promotion of mental health'.[26] Bowlby's reference to US opinion is intriguing for by this point Benjamin Spock's *Baby and Child Care* was becoming the key text in post-war American child-rearing. As Kathleen Jones suggests, the book's success 'attested to the popularity of the child guidance and mental hygiene movements and the success of [their] popularizers in reshaping the discourse of troublesome childhood'.[27]

In the lay publication *Mother and Child* Bowlby tackled the issue of the 'bad' parent. It was a basic premise 'that, for good or bad, parents are indispensable! It is sometimes forgotten that they are as necessary a part of the child's environment as fresh air and good food.' Of 'bad' parents, though, criticism was pointless. Rather, it was necessary to 'understand the bad parent in order to help her [*sic*]'. What also had to be taken into account was that 'behind most bad parents there lies an unhappy childhood'.[28] So, as Bowlby put it on another occasion, bad parenting was 'best looked upon not as something to be criticised, but as an illness to be studied'. Those 'bad' parents who had apparently given up on their responsibilities only showed in their 'defeatism' how far their illness had progressed, often aided by the 'unhelpful moralizing of the community'.[29] For Bowlby parents might well fail but, given his views of institutions encountered earlier, how much greater was society's failure if their children had to be removed to such places?

[26] J. Bowlby (1953) *Child Care and the Growth of Love* (Harmondsworth: Penguin Books), pp. 101–2 and *passim*.

[27] K. Jones (1999) *Taming the Troublesome Child: American Families, Child Guidance, and the Limits of Psychiatric Authority* (Cambridge, MA: Harvard University Press), p. 117.

[28] J. Bowlby (1950) 'The Parents' Influence for Good or Bad', *Mother and Child*, 21:5, 135–36.

[29] J. Bowlby (1953) *The Roots of Parenthood* (London: National Children's Home), p. 16.

Bowlby and other child guidance practitioners viewed themselves as non-judgemental and working in a scientific manner. For them bad parenting, like childhood maladjustment, was a pathological condition and not the result of moral failing. It could thus, like any other 'illness', be cured through appropriate intervention. This was hugely preferable to placing children in rigid and uncaring institutions or taking them away from their parents in some other way, hence the didactic articles in popular outlets such as *Mother and Child*. Another way of tackling potential maladjustment was outlined by the psychiatrist and medical director of a child guidance clinic, Helen Gillespie, who reported that her survey of 100 pre-school children had revealed that 90 per cent were presenting problems. Maladjustment was thus in place before any contact with the school medical service so Gillespie had inaugurated a clinic at a maternity and child welfare centre for infants. Children at this age might suffer from a range of issues which Gillespie discussed with the mother to assess the latter's attitude to her child's problems, her own anxieties, her acceptance of the child, and any tendency to transfer her own psychological problems onto the child.[30] Attitudinal improvement in the mother would therefore result in behavioural improvement in her offspring.

For the Seebohm Committee, meanwhile, there were underlying principles which should guide child welfare policies. It specifically felt that the ultimate responsibility for children lay with their parents. This was contrasted with previous practices when state authorities had countered neglect, for example, by taking full parental rights and removing children from their homes. Parents' duties notwithstanding, though, there was a need for a sharing of responsibility between parents and the state and for a conscious policy that wherever possible children should be treated in the context of their families, communities, and schools. The Committee believed too that, again whenever possible, help should be on the basis of 'a genuinely voluntary agreement between parent and authority'.[31] More broadly, Welshman suggests, the Committee was 'the first serious examination of health and welfare services' since the post-war legislation noting, *inter alia*, psychiatrists' increasing support for community care alongside a

[30] H. Gillespie (1955) 'Psychiatric Problems of Children Under Five Years', *Mother and Child*, 25:11, 272–74.

[31] Secretary of State for the Home Department et al., *Report of the Committee on Local Authority and Allied Personal Services*, pp. 56–58.

patchy and limited provision of such services.[32] Again, this was part of the perceived need to shift away from institutional solutions to one attuned to individuals and their families.

Those who were going to carry out this support of children and their families were members of the so-called 'psy' professions, in turn part of the rise of the expert and the psychologising of society. Psychiatrists, psychologists and psychiatric social workers had the shared aim of establishing professional identity and autonomy. This was to be achieved preferably and increasingly outside of institutions such as asylums or children's homes. What was therefore stressed was their professional roles in individual, and individualised, clinical practice, and in the case of child guidance in a dedicated but non-residential setting. Each child was unique and had to be treated as such. Institutions were not the place to do this—on the contrary. Psychiatric social workers were particularly significant given their place on child guidance's front line. As a result of their training, and from the context in which they operated, psychiatric social workers saw their focus as being on the child/parent relationship and on supporting this by way of parental education in the home or the clinic. Such beliefs had been strongly influenced by American psychiatric social work which, as in Britain, had played a foundational role in child guidance. As a result of the high status that psychiatric social work held within the profession as a whole this became a dominating approach in social intervention.[33]

What other implications did such ways of seeing childhood psychological and emotional problems have? We have seen that child guidance practitioners claimed to be non-judgemental and their views based on science. But it was more complicated than this. Whatever the practitioners' intentions, their belief that childhood was inherently problematic—something could go wrong at any time and if it did it was almost certainly the fault of bad parenting—must have induced considerable anxiety amongst parents themselves. Child guidance was also predicated on a particular model of the family, its members' functions, and what constituted normal behaviour, all, it might be argued, socially and historically constructed rather than fixed and unchanging. Here Bowlby again illustrates these points. In an NAMH pamphlet of the late 1950s entitled *Can I Leave My Baby?* he argued that being a mother was 'a craftsman's job and perhaps the most skilled in the world'. But it was an 'exacting job' and 'scamped at one's

[32] Welshman, 'Rhetoric and Reality', p. 213.
[33] See further Stewart, *Child Guidance*.

peril'. It was not possible to 'really give back to a child the love and attention he needed and did not receive when he was small. With understanding and affection, and perhaps skilled help, one can go a long way towards it, sometimes a very long way, but it will never quite be the same.' Even more specifically, when the mother is absent various others could stand in temporarily, including fathers, as 'spare Mummy'. Nonetheless for small children it was their mothers who were most important to them so, during these early years, 'most fathers come a poor second'.[34]

The centrality of the family was endlessly stressed and became the post-war orthodoxy. Hendrick cites a Home Office circular of the 1940s which argued that: 'To keep the family together must be the first aim, and the separation of a child from its parents can only be justified when there is no possibility of securing adequate care of a child in his own home'.[35] Two British psychiatric social workers who attended the 1955 World Child Welfare Conference found the most impressive speech that of the professor of psychology at the University of Strasbourg. He described the fundamental attitudinal change over the last half century, 'from the ready placement of deprived children in large institutions to a realisation of the need to keep its members together if at all possible, or, in the event of complete breakdown, to find some kind of substitute family'. Among the authorities cited by the speaker on the 'perils of early separation' was Bowlby.[36] Here is further evidence that scepticism about institutionalisation was not confined to Britain and that Bowlby's work had more than a domestic impact.

A few years later the 1959 Mental Health Act sought to move mental healthcare provision into the community and services such as child guidance can be seen as in the vanguard of this shift. Community care was given further, controversial, impetus through a speech given to an NAMH conference by Enoch Powell, the Minister of Health, in 1961. Powell signalled his intention to close down large-scale mental hospitals. It was not only children who were to be deinstitutionalised. The same meeting was rather also addressed by Professor of Child Health, A. V. Neale. Child guidance was, he claimed, indicative of 'new trends in medico-psychological

[34] J. Bowlby (1958) *Can I Leave My Baby?* (London: National Association for Mental Health), pp. 6–7.

[35] Hendrick, *Child Welfare*, p. 139.

[36] K. Mawhood and I. Elkan (1955) 'World Child Welfare Congress, Zagreb', *British Journal of Psychiatric Social Work*, 3:1, 37–38.

aims' which could be summarised as doctors having an educational role in preventive medicine. Infant welfare had seen a shift from nutritional matters to psychological analysis of the mother/child relationship. A child's early years created the 'capital for the mental health of later years'. The 'natural history of the affectionless child' led to antisocial attitudes and delinquency in later life. Emotional disturbances on the part of parents, furthermore, led to problems for their children. And since every child was unique each had to be dealt with and treated individually and holistically.[37] Once more we see medical and individualised diagnoses and treatments preferred to those which stressed moral failure and non-familial solutions to childhood maladjustment.

Constraints and Challenges

The child guidance approach did not, however, go unchallenged. Child guidance had, throughout its history, its critics and indeed in the post-war era was beset by bouts of self-criticism and self doubt.[38] As Hendrick remarks, even when the 1948 Children Act was being introduced there were those who questioned whether leaving children with their families was always a wise or safe approach.[39] Nonetheless, the primacy of the family in child-rearing was widely accepted.

What, though, were the actual outcomes and how did the child guidance message come to be challenged? As we have seen, supporters of deinstitutionalisation based their arguments primarily on psychological grounds while also noting the cost benefits. However none of this meant that all residential institutions for children disappeared overnight, or that child guidance and other community psychiatric services became the norm. The relative paucity of investment in welfare, including social services, meant that for children deinstitutionalisation was a slow process while initiatives dealing with mental health at whatever level were notably ill-served. The Seebohm Committee observed that one of modern psychiatry's main discoveries was the recognition of the 'profound influence' of social environment on mental well-being. Nonetheless, there remained no satisfactory way of estimating the number of people, of whatever age,

[37] A. V. Neale (1961) 'Integration' in National Association for Mental Health, *Emerging Patterns for the Mental Health Services and the Public* (London: NAMH), pp. 13, 18.

[38] Stewart, *Child Guidance*, *passim*.

[39] Hendrick, *Child Welfare*, p. 138.

who came into contact with the mental health services. And, the 1959 Act's aspirations notwithstanding, the idea that there existed comprehensive community-based facilities for the mentally ill was a 'sad illusion'.[40] In terms of numbers of children in various forms of local authority care, direct comparison between the data provided by Curtis and by Seebohm is problematic because of changes in the law, provision, and methods of classification. But the former showed that in the mid-1940s just over 40,000 children were under local authority care plus nearly 28,000 in foster care and 40,000 in voluntary homes and hostels.[41] At the end of the 1960s Seebohm recorded 69,000 children in all forms of local authority care.[42] There was clearly still a long way to go if every child was to be brought up in a family environment. Indeed Thomson points to a subsequent rise in children in local authority care through to the mid-1970s.[43]

Broader social and social welfare change was, moreover, in the offing. These changes did not necessarily question deinstitutionalisation, but some did begin to undermine one of its foundations, child guidance. First, the 'rediscovery of poverty' of the 1960s, led by voluntary campaigning bodies such as the Child Poverty Action Group, suggested that socioeconomic disadvantage was an important factor in family stress and breakdown, claims to the contrary notwithstanding. This was also a challenge to the mildly social democratic consensus of the post-war era, of which child guidance was a part, in that it suggested that the welfare state had failed to deliver its promise. But an ultimately much greater threat to the social democratic consensus was its own disintegration in the face of sluggish economic growth, external economic shocks and the rise of the New Right. The New Right was hostile to the welfare state stressing the need for individual responsibility rather than any focus on purported social or familial pathology.[44]

Second, as noted, child guidance had always had its critics and in the post-war period some social scientists, for example the influential Barbara

[40] Secretary of State for the Home Department et al., *Report of the Committee on Local Authority and Allied Personal Services*, pp. 107–108.

[41] The Home Department and the Ministry of Education, *Report of the Care of Children Committee*, p. 27, Table IV.

[42] Secretary of State for the Home Department et al., *Report of the Committee on Local Authority and Allied Personal Services*, p. 52.

[43] Thomson, *Lost Freedom*, p. 93.

[44] For child welfare policy under the Conservatives in the late twentieth century, Hendrick, *Child Welfare*, ch. 5.

Wootton, were notably sceptical about, in particular, its emphasis on maladjustment's psychological origins.[45] And as Thomson observes, another eminent social scientist, the anthropologist Margaret Mead, argued that there could be more than one model of child-rearing as evidence from other cultures demonstrated.[46] Attacks from these quarters challenged, *inter alia*, child guidance's claims to scientific status as well as its operational methods. More broadly, feminists were beginning to question the ideological basis of the 'traditional' family and women's role therein.

Third, the Seebohm Committee's primary function was to investigate the organisation of personal social services. Among its recommendations were that all social work functions be focused in local authority departments and that social work training be 'generic' rather than specialised. The aim here was that a family coming into contact with the social services would be dealt with by one social worker who would address all their needs rather than by, potentially, a mixture of specialists and generalists from a range of agencies.[47] For present purposes what is important is that, along with the reorganisation of the NHS in 1974, this effectively ended the profession of psychiatric social worker. Given the prominence of psychiatric social workers in child guidance, and their fundamental belief in parental responsibility for child maladjustment, this was a blow both to an approach which focused on familial psychology and to a previously dominant form of social work practice.

Fourth, new views on 'problem families' were emerging. The notion of the 'problem family' has a long pedigree. But the fresh development of the early 1970s was the concept of the cycle of deprivation, controversially articulated by the leading Conservative politician and early neoliberal, Sir Keith Joseph. As Welshman notes, Joseph argued that there existed a dynamic process wherein familial dysfunction was transmitted down the generations so depriving individuals of full physical, emotional and intellectual development; and wherein poverty was one of a number of possible causal factors. Joseph also appeared to imply, in a manner which critics saw as dangerously close to eugenics, that transmitted deprivation had a hereditary component and that some families of themselves were not necessarily enough to provide an adequate culture for child-rearing.

[45] Stewart, *Child Guidance*, p. 183.

[46] Thomson, *Lost Freedom*, p. 94.

[47] Secretary of State for the Home Department et al., *Report of the Committee on Local Authority and Allied Personal Services*, chs 12, 13.

There were clearly some areas of overlap here with child guidance, notably in the idea that problems could be passed down the generations, but there were also significant differences. Child guidance maintained that all children were potentially at risk and that what was needed was psychological guidance for parents. Among the proposals of proponents of the cycle of deprivation were that there be better provision in areas such as pre-school playgroups and nurseries and that women in certain categories be encouraged to use contraception. They also believed that certain types of family were especially troubled and thereby more inclined to pass problems generationally.[48] Unlike child guidance, then, there was no perceived need to target the whole child population.

From a different perspective, some child guidance practitioners were beginning to acknowledge causes other than bad parenting in childhood maladjustment. At a 1967 conference one speaker suggested the need to deal with families who did not attend a child guidance clinic but who nonetheless had been brought to the clinic's attention. She sought to distinguish these 'disorganised families' from 'problem families' while acknowledging that they shared certain characteristics. Disorganised families were not only emotionally dysfunctional but also poor, large, viewed unfavourably by their communities, and already known to a number of agencies. Hence if child guidance 'is to have a place in the community then it must share the community's concerns' in what was now 'the age of community care'.[49] Although coming from very different positions, what is notable about both approaches just described is that, while breaking up the family is not advocated, each aimed to target particular families while rejecting the notion that dysfunction was solely attributable to the parent/child relationship. This too began to call into question child guidance's underlying principles.

Finally, there was growing unease with the family as unquestionably the location for child-rearing. This was fuelled by the 'discovery' of the 'battered baby' and other forms of abuse, including sexual abuse, within family settings. As Hendrick observes, the 1973 murder of 6-year-old Maria Colwell by her stepfather in the family home had a profound impact on both public opinion and official thinking about child welfare. Among the

[48] J. Welshman (2013) *Underclass: A History of the Excluded*, 2nd edn (London: Bloomsbury Academic), ch. 6.

[49] Wellcome, SAMIN/B/64, Miss P. Parsloe (1967) 'Working with Families Who Do Not Attend a Clinic' in National Association for Mental Health, *Child Guidance from Within: Reactions to New Pressures: Papers Given at the 23rd Child Guidance Inter-Clinic Conference, 1967* (London: NAMH), pp. 41–52.

case's many disturbing aspects was that the child had spent over five years in the foster care of her aunt and had not wanted to return to the family home. The social services department, though, had 'a long term plan to reintegrate Maria with her natural mother's family's family'. It also became clear that the child, the department's aspirations notwithstanding, had had 'a weak or non-existent bond with the natural mother'.[50] The reaction to the case was complicated. But the Colwell case certainly cast doubt on the naturalness and centrality of the mother/child bond and on the idea that bad parents were better than none and in any event could be re-educated.

Conclusion

Child guidance's practice and philosophy were central to arguments for the deinstitutionalisation of childcare in the quarter century after the Second World War. Such views were articulated in a range of outlets by influential individuals, organisations and official committees. These arguments gained purchase partly because of Bowlby's growing influence, his emphasis on the importance of the mother/child bond, and his hostility to institutionalisation. But even before Bowlby rose to prominence child guidance had shown itself in favour of family-based solutions to child maladjustment, marking its early post-war impact by way of the Curtis Committee, the 1948 Children Act, and its own place in the welfare state before going on to play a powerful part in shaping child welfare policy in the ensuing quarter century. It is also worth noting what a tightly-knit group, and one which generally spoke with one voice, we have encountered. Bowlby, for example, was an early member of the NAMH Council, a frequent speaker at its meetings, and at one point president of the Association of Psychiatric Social Workers. The broader context is also important here with, for example, the emphasis on community-based mental health policies in the 1959 Mental Health Act. However, it has to be acknowledged that child guidance's principles were increasingly called into question by, for instance, those who pointed to socioeconomic factors in child and family stress and dysfunction. While such criticisms far from constituted a positive argument for institutional care they nonetheless called into question the notion of the family as, in most circumstances, the best place for child-rearing and subsequent policies sought to prioritise the child's interests above all else.

[50] Hendrick, *Child Welfare*, pp. 163–64.

CHAPTER 10

'Do You Have a Frog to Guide You?': Exploring the 'Asylum' Spaces of R. D. Laing

Cheryl McGeachan

Introduction

'Do you have a frog to guide you?' Stuart's voice was impervious, demanding.
'Huh?' Startled, I turned to face him. From beneath his soft wide-brimmed hat the suave roué frowned at me like an aroused customs officer. 'What did you say?'
'Do-you-have-a-frog?' He fixed me with a pale stare, the schoolmaster impatiently tending his dim-witted pupil. ...
'I'm not quite sure what you mean by that', I hesitated. This was the first time Stuart had spoken to me, and only my second visit to Kingsley Hall. I was still scared of what the crazies might do. I half expected him to reach into his pocket and produce a croaking bullfrog.[1]

[1] Mike (undated) 'A Personal Memoir' (Unpublished Typescript, Papers of R. D. Laing, Special Collections, University of Glasgow: MS Laing A711), p. 2. All names in this memoir are pseudonyms, except from the author, in order to protect the identity of the individuals

C. McGeachan (✉)
University of Glasgow, Glasgow, UK
e-mail: Cheryl.McGeachan@glasgow.ac.uk

D. Kritsotaki et al. (eds.), *Deinstitutionalisation and After*,
Mental Health in Historical Perspective,
DOI 10.1007/978-3-319-45360-6_10

This opening quotation comes from a personal memoir of a resident in one of R. D. Laing's 'asylum' spaces, Kingsley Hall, developed in the 1960s by the Philadelphia Association (PA) as an alternative to the large-scale institutionalisation process of the past centuries. Stuart was one of many individuals in the period attempting to find a place of safety and refuge during times of extreme mental difficulty. However, as the quotation demonstrates, this was not an easy process, with fear and miscommunication happening regularly between people in these dark and challenging spaces.[2] Attempts to create places of so-called true asylum have continued to haunt a range of psychiatric professionals throughout the centuries and this chapter seeks to turn the attention to R. D. Laing and his colleagues in the 1960s, in order to highlight the range of asylum spaces they envisioned and created in the wider changing post-asylum landscape in Britain during the period.

'Madness', as Philo has noted, has a complex geography and many geographers have attempted to navigate their way through its difficult terrains.[3] While some have sought to investigate the spatial shifts associated with the history of madness (and subsequently asylums and community care as discussed below), others have turned to focus on the lived experience and lived geographies of individuals experiencing mental ill-health.[4] For Parr:

> Attention to the lived experiences of those with mental health problems historically in the asylum and contemporarily 'on the streets', is not merely to correct the methodological imbalance, but necessary in order to articulate how the story of madness and illness is not simply, or just one of exclusion, subjectification and outsiderness. Instead, people with mental health problems should also be understood as creative actors, often capable of resistance, self- and collective empowerment and determination in the diverse spacings of madness, illness and mental health care.[5]

referred to. This is due to the material not being publicly available and only accessible via the R. D. Laing Collection held in Special Collections, University of Glasgow.

[2] In the above mentioned memoir the author discusses in detail the dark atmosphere of the house.

[3] C. Philo (2004) *A Geographical History of Institutional Provision for the Insane from Medieval Times to the 1860s in England and Wales: 'The Space Reserved for Insanity'* (Lampeter: Edwin Mellen).

[4] H. Parr (2008) *Mental Health and Social Space: Towards Inclusionary Geographies?* (Oxford: Blackwell).

[5] Ibid., 12.

When discussing individuals such as Stuart, caught up in a myriad of complex worlds and social spaces, there is arguably a need to engage with the everyday geographies of these experiences. In discussing Laing's ability to 'world' his psychiatric and psychoanalytic insights, I have previously discussed his desire to pay acute attention to the life-worlds of his severely disturbed patients.[6] Through Laing's time spent in a range of psychiatric institutions he became increasingly aware that individuals must somehow 'live on' in, and through, the many different societies, institutions and family environments which they inhabit and therefore it becomes important to pay specific, significant attention to the wide range of these individual and collective spaces and places.

Alongside these attempts to chart individual experience is an attention to the sites and spaces that have been used to house those experiencing mental health problems across the centuries. For many geographers the figure of the asylum looms heavily on the landscape and becomes a critical site for investigation. Arguably, acute attention to the space(s) of the asylum can cast light on the complex geographies *of* madness and societal responses *to* it.[7] Questions raised in this type of work address the extent to which such locations—the specific spaces and environments—were shaped by medical, moral or economic discourses, or by a deeper sense of wishing to remove certain 'troubling' and 'frightening' populations from 'sight and mind'.[8] Many of these arguments have been debated in reference to a large number of similar institutions, not specifically designed to house and treat those with mental health problems but which inevitably become spaces and places of care and incarceration for such populations, including prisons and orphanages.[9] Studies into such places reveal the complexity of mental healthcare across the centuries and lay the foundations for questioning such asylum spaces as strategies for dealing with individuals experiencing mental health problems which began to diversify in the contemporary landscape.

[6] C. McGeachan (2014) 'Worlding Psychoanalytic Insights: Unpicking R. D. Laing's Geographies' in P. Kingsbury and S. Pile (eds) *Psychoanalytic Geographies* (Farnham: Ashgate).

[7] Philo, *Geographical History*.

[8] Ibid.

[9] D. Moran (2015) *Carceral Geography: Spaces and Practices of Incarceration* (Farnham: Ashgate); T. Disney (2015) 'Complex Spaces of Orphan Care—a Russian Therapeutic Children's Community', *Children's Geographies*, 13, 30–43.

Alongside these institutional spaces and their legacies, geographers have begun to examine a wide assortment of care spaces in the community that emerged out of, or in response to, the process of deinstitutionalisation, such as outpatient clinics, hostels, support projects, counselling services and home spaces.[10] While some have expressed concerns about the sufficiency of such spaces in terms of numbers and quality of service, others have questioned whether the new mechanisms of 'community care', dispersed throughout diverse communities, actually amount to a tighter web of 'psychiatric influence'.[11] For many the demise of the asylum structure has created a new system of segregation for the ex-mental patient becoming apparent in the contemporary landscape.[12] As the old asylum buildings are being destroyed or redeveloped difficult questions remain about their influence and legacies on mental healthcare in a deinstitutionalised, or 'post asylum' landscape that remains under-examined.[13] While studies on asylums and their patients will long continue, I seek to argue that attention needs to turn now to the alternative spaces of asylum and the individual lived geographies of mental ill-health that they may reveal. In doing so, further insight and understanding of individuals such as Stuart and the complex landscapes of deinstitutionalisation they are wrapped up within will be highlighted.

Using the example of Scottish psychiatrist R. D. Laing and his colleagues, such as Aaron Esterson, who practiced during the 1950s and 1960s and became influential for changing and developing a new landscape of mental health care in Britain, this chapter seeks to argue that alternative conceptualisations of therapeutic asylum spaces (both real and material, imagined and envisioned) matter in the exploration of the deinstitutionalisation process. Through investigating Laing's published and unpublished work this chapter aims to investigate the spatial elements to his design and implementation of experimental therapeutic pursuits at a time of significant social and political change. Concentrating specifically on Laing's time spent at the Tavistock Institute of Human Relations in

[10] Parr, *Mental Health*; L. Bondi (2009) 'Counselling in Rural Scotland: Care, Proximity and Trust', *Gender, Place and Culture*, 16, 163–179; L. Bondi and J. Fewell (2003) '"Unlocking the Cage Door": The Spatiality of Counselling', *Social & Cultural Geography*, 4, 527–547.

[11] J. Wolpert and E. Wolpert (1974) 'From asylum to ghetto', *Antipode*, 6, 63–76.

[12] Parr, *Mental Health*.

[13] G. Moon, R. Kearns and A. Joseph (2015) *The Afterlives of the Psychiatric Asylum: Recycling Concepts, Sites and Memories* (Farnham: Ashgate).

London and his part in the development of the therapeutic communities of the PA during the early 1960s, this piece seeks to unravel a further dimension to his controversial studies on mental health that remains under-explored. Beginning with a discussion of Laing's own blueprint for a true asylum this chapter will then examine Laing and Esterson's investigations into familial worlds. Finally, through exploring the therapeutic pursuits of the PA, including Kingsley Hall and the Archway Community, this chapter demonstrates the importance of investigating alternative spaces of asylum in the deinstitutionalised landscape, from their very imagining to their physical foundations.

Laing's Asylum and Post-Asylum Thinking

> I want to ponder a little over the idea of asylum, in the most general, widest, and profoundest sense.[14]

The Scottish psychiatrist, psychoanalyst and psychotherapist Ronald David Laing, born in Glasgow in 1927, became globally renowned for his work with severely psychotic individuals. Throughout his life he wrote numerous works exploring the process of going 'mad' and contemplating the possibility of the social intelligibility behind it. Laing's work, particularly his early texts, *The Divided Self*, *The Self and Others*, *Self and Others* and *Sanity and Madness and the Family*, move people in a range of ways and the legacy of his work is seen in many guises.[15] Laing's insights uncover a host of possibilities to connect the stories of individuals suffering from mental health problems to a wider spectrum of literature concerned with understanding people in the context of their own life-worlds. Laing's interest in these individuals was greatly attached to *where* he encountered them, from within the confines of the mental hospital to the private spaces of their family homes. Throughout his early career Laing was concerned

[14] R. D. Laing (1985) 'The What and Why of Asylum', Unpublished Typescript, Papers of R. D. Laing, Special Collections, University of Glasgow (MS Laing A532), p. 1.

[15] R. D. Laing (*c.*1960/1990) *The Divided Self: An Existential Study in Sanity and Madness* (England: Penguin Books); R. D. Laing (1961) *The Self and Others: Further Studies in Sanity and Madness* (Tavistock: Tavistock Publications); R. D. Laing (1969/1972) *Self and Others* (Harmondsworth: Penguin Books); R. D. Laing and A. Esterson (1964/1990) *Sanity, Madness and Family: Families of Schizophrenics* (Harmondsworth: Penguin Books); L. Fowler (2011) *All Divided Selves* [film] (Gisela Captain Gallery: The Modern Institute/Toby Webster Ltd).

with the spaces and places of care and began to manipulate therapeutic spaces in the institutions where he worked, starting to imagine and create his own versions of caring places. He came to believe strongly that many of the places formed to look after these individuals were not of a therapeutic nature.[16] This attention to the spatial details of his patients' worlds and the changing landscapes of care makes him an intriguing figure for geographical investigation into the deinstitutionalisation process.

From his previous time working in Army psychiatric hospitals and NHS mental institutions across the UK during the 1950s, it appeared that Laing was becoming increasingly disillusioned as to how the therapy that he was seeing, and delivering, could ever be effective in binding his ideas together with the changing climate of institutionalisation in Britain at the time.[17] From the 1950s the UK government sought to change service provision in mental health significantly through the policy termed 'care in the community'. A key aspect of this policy was to change the pattern of care for these individuals from grand institutional inpatient facilities—traditionally, asylums then mental hospitals—to a wide range of community-based services. Whilst Laing worked at Gartnavel Royal Hospital in Glasgow during the 1950s, the UK saw its peak of psychiatric service beds and he witnessed first-hand the difficulties faced by these struggling institutions. However, as Parr notes, 'community care became a watchword of mental health services in the 1970s' and the first full hospital closure did not occur until 1986.[18] Therefore, Laing's experience of mental healthcare came during a significant transitional phase that led to important challenges to established medical and political order, and fierce debates surrounding the ways in which individuals diagnosed as mentally ill should be viewed and treated.

For Laing, and also others involved in the psychiatric and psychotherapeutic professions such as Aaron Esterson and David Cooper,[19] the idea

[16] J. Andrews (1998) 'R. D. Laing in Scotland: Facts and Fictions of the "Rumpus Room" and Interpersonal Psychiatry' in M. Gijswijt-Hofstra and R. Porter (eds) *Cultures of Psychiatry and Mental health Care in Postwar Britain and the Netherlands* (Amsterdam: Rodopi); C. McGeachan (2013) 'Needles, Picks and an Intern Named Laing: Exploring the Psychiatric Spaces of Army Life', *Journal of Historical Geography*, 40, 67–78.

[17] McGeachan, 'Needles, Picks and an Intern Named Laing'.

[18] Parr, *Mental Health*, 32.

[19] Cooper is an important voice in the range of movements during the 1960s and 1970s to create alternative forms of asylum.

of therapy was tied too tightly to the mental institution, and it was this connection that Laing and others desired to change. Laing declared:

> I would like to see mental hospitals as more the sort of sanctuary that I think they should be. I think mental hospitals ought to be secure places of refuge, in other words, genuine asylums, but mental hospitals we have today ... have confused their function, I think, with their presence and have confused their function/relationship to society with that of—something like a mechanic where the idea is to re-service people who are thought of as having broken down in some mechanical behavioural way—to re-service them as quickly [and] as efficiently as possible to get them back functioning in society.[20]

Insistent on believing that 'mad or not, a human being is first and foremost a human being', Laing sought to break down the barriers of conventional thinking about the way in which therapy could be conducted through his notion of 'true' asylum.[21] Questions of asylum were consistently reflected upon by Laing throughout his life and in a typescript housed in his archive entitled 'The What and Why of Asylum', he sketches out his vision for future spaces of provision for those experiencing mental distress.[22]

Discussing the abolition of the old mental hospital system, for example, the reforms occurring in Italy promoted by Franco Basaglia, Laing questions the features of these incarceration procedures that have actually been eradicated, suggesting that careful reflection must be given to the new abuses that may be present in these freshly implemented systems. In doing so, Laing stresses he is 'trying to get into focus the idea of a place where safety and security is the primary intent and end in itself' and calls for the creation of a 'small-scale strategy' for genuine asylum care.[23] For Laing this small-scale strategy of asylums, sanctuaries and group homes set up by voluntary organisations and local authorities was already being implemented quietly across in the UK by the 1980s, allowing the creation of

[20] Laing as quoted in G. M. Carstairs (1973) 'Something To Say No. 241/2 John Morgan Speaking to R. D. Laing', Papers of R. D. Laing, Special Collections, University of Glasgow (MS Laing L149), p. 3.

[21] J. Wykert (1978) 'R. D. Laing at 50', *Psychiatric News*, 32.

[22] Multiple copies of these notes have been typed up and edited by Laing over the years.

[23] Laing, 'The What and Why of Asylum', p. 6.

safe and secure places for frightened and insecure people to access in times of need.[24]

These notes, sketched out and edited throughout many different periods of his life, denote a conviction to find an alternative spatial solution in society—at an alternative scale—to the old psychiatric system erected to contain individuals in grand asylums. However, importantly, this was not a call to abandon psychiatric care for those who required assistance, as is frequently cited by anti-psychiatrists, but an attempt to find a range of ways to provide a 'life support system' for the most vulnerable and desperate in society.[25]

Laing's idea of asylum as, therefore, a place of safety, 'a time and place in which one *is*', relates to the deep importance that he placed on the environment as a factor in both health and illness.[26] Indeed, he wrote as follows:

> Remember our environment does not merely *environ* us. It does not only surround us and contain us. It is inside us ... We are more than drenched by the world, more than we are sodden with it.[27]

It is this key role of environment in the making and shaping of selves that led Laing to think more carefully about the combination of therapy with the precise types of places in which different, especially alternative, forms of therapy could be carried out. For Laing, the original meaning of therapy was 'attendant', meaning that a therapist is someone who attends to another's needs, a notably difficult task.[28] He spoke of developing a four-pillar therapeutic approach which would contain an imperative to work with the bodily, mental, emotional and social phenomena of the situation. This would allow the patient and the therapist to adopt a stance that did not place one as either inferior or superior, but on the same level:

[24] Ibid., p. 4.

[25] Ibid., p. 2.

[26] Laing as quoted in D. Glotzer (1973) 'Experiencing R. D. Laing', *Valley Advocate*, p. 9, emphasis in original.

[27] R. D. Laing (1974) 'Draft notes', Papers of R. D. Laing, Special Collections, University of Glasgow (MS Laing A682), n.p., emphasis in original.

[28] R. D. Laing (1974) 'Transcripts of interviews with R. D. Laing and John McGreevy', Papers of R. D. Laing, Special Collections, University of Glasgow (MS Laing L162/1–5), p. 22.

'a stance of "humility", if the word could be truly understood'.[29] His vision of asylum, therefore, could only be realised in conjunction with the notion of human solidarity at the very centre of issues relating to the environment and therapy itself. Throughout his published work and his social experiments, Laing showed a conviction that relationships can heal and that communication, occurring through a range of sensory channels, was bound unbreakably to the places and therefore environments within which it was to be conducted. The following section will explore these issues in more detail through Laing's work with fellow Scottish psychiatrist Aaron Esterson and their work on the family.

Sanity, Madness and the Family

Throughout Laing's time working and training at the Tavistock Clinic and the Tavistock Institute of Human Relations in the late 1950s and early 1960s, there was always an element of frustration within his work as he felt he was being dragged away from working with the long-term so-called psychotic individual to dealing with outpatients or so-called neurotics. Laing's passion was the plight of the persons termed 'psychotic' or 'schizophrenic', where he described the literal meaning not to be 'split mind' or 'split personality' but a condition of broken-heartedness.[30] Teaming up with fellow psychiatrist Aaron Esterson,[31] who held a number of appointments in British mental hospitals and psychiatric units, they undertook research in which psychiatric patients labelled as schizophrenic could be evaluated within their home environments and in a range of

[29] Ibid., n.p.

[30] R. D. Laing, A. R. Lee and D. Sherret (1960) 'Draft Application for a Grant for the Study of Schizophrenic Patients and their Family Interactions, Tavistock Clinic', Papers of R. D. Laing, Special Collections, University of Glasgow (MS Laing A620), p. 2.

[31] Esterson was born in Glasgow in 1923 and, after being demobilised from the Royal Navy, entered Glasgow University where he graduated as a doctor of medicine in 1951. This is also where he first met Laing and their friendship would continue until coming to a bitter end after the setting up of Kingsley Hall in 1965. Esterson worked in general practice until taking up psychiatry in 1954, and between then and 1962 he held a number of appointments in British mental hospitals and psychiatric units. Esterson became disenchanted with the practice of therapy in the National Health Service and he entered private practice in 1962 as an existential psychoanalyst and family therapist.

family scenarios with the hope of forming a greater understanding of their whole situation and the types of personal assistance they required.[32]

The controversial study, produced over a period of five years, was published as *Sanity, Madness and the Family: Families of Schizophrenics* and consisted, in many ways quite simply, of 11 case studies of female patients and the stories of their lives and families as presented to the researchers.[33] These women were all diagnosed schizophrenic by at least two senior psychiatrists and had all been admitted to hospital, some for a short period of time and others for a prolonged period.[34] However, it is important to note that, as psychiatrists and analysts, Laing and Esterson did *not* accept schizophrenia as a biochemical, neurophysiological or psychological fact, signalling their desire for this research to be read not as a study of schizophrenic patients but as people who had been diagnosed schizophrenic by a medical source. The interest in this study was on relations and networks between people in different familial formations and it connected extensively to the wider network of studies focusing upon families and schizophrenia of the period.[35] It was clear to both men that to see the patient in isolation from their social situation, their immediate, everyday social environment, in this case their family homes, was to miss an important interpersonal aspect of the individual's world. Thus, the author's note:

> Each person not only is an object in the world with others but in a position in space and time from which he experiences, constitutes, and acts in *his* world. He is his own centre, with his own point of view, and it is precisely each person's *perspective* on the situation that he shares with others that we wish to discover.[36]

This concentration on time–space situatedness is intimately connected to the existential and phenomenological traditions from which they both drew considerable inspiration and has clear geographical and spatial res-

[32] Although both Laing and Esterson wished to conduct the interviews in the home, the majority of the interviews actually took place within the institutions themselves.

[33] This study received public criticism with many taking the prominent argument to be that 'families cause schizophrenia'. A. Laing (2006) *R. D. Laing: A Life* (Stroud: Sutton Publishing).

[34] Laing and Esterson, *Sanity, Madness and the Family*, p. 15.

[35] J. Neill (1990) 'Whatever Became of the Schizophrenogenic Mother?', *American Journal of Psychotherapy*, 44, 499–505.

[36] Laing and Esterson, *Sanity, Madness and the Family*, p. 19, emphasis in original.

onances that reach across to the wider consideration of deinstitutional approaches for dealing with individuals experiencing mental ill-health.[37]

Laing and Esterson had to create a methodology that would work for such a project and discussed this at length in their introduction to the text.[38] Not only were they attempting to engage in an assessment of the intricate behaviour that existed between family members in a variety of combinations, but they were attempting also to move away from the artificial setting of the mental institution into the place from which the person originally came: the home. The home environment was hence a key factor for both Laing and Esterson, and they were acutely aware of how important this space was for the people they were treating. The introduction to the study insisted:

> The way in which a family deploys itself in space and time, what space, what time, and what things are private and shared, and by whom—these and many other questions are best answered by seeing what sort of world the family has itself fleshed out for itself, both as a whole and differentially for each of its members.[39]

From this perspective, Laing and Esterson attempted to frame a clear research aim for their project:

> Our question is: are the experience and behaviour that psychiatrists take as signs and symptoms of schizophrenia more socially intelligible than has come to be supposed?[40]

In seeking out social intelligibility of the schizoid individual they wished to present the family's thoughts and words that were uncovered from inside the texture of the home itself, giving a sense of not only the family's thoughts and experiences but a whole range of elements that made up their worlds.[41] This family research contained important notions about tracing schizophrenia *out* of the traditional asylum environment and *within* a host

[37] C. McGeachan (2014) '"The World Is Full of Big Bad Wolves": Investigating the Experimental Therapeutic Spaces of R. D. Laing and Aaron Esterson', *History of Psychiatry*, 25, 283–298.

[38] Laing and Esterson, *Sanity, Madness and the Family*, pp. 15–27.

[39] Ibid., p. 21.

[40] Ibid., p. 12.

[41] For further discussion of the family case studies used in the text, see McGeachan, '"The World Is Full of Big Bad Wolves"'.

of different spaces, and this contributed greatly to Laing and Esterson's 'post-asylum' vision. While the antipsychiatry movement gained momentum in this period, Laing and Esterson sought not to demonise the space of asylum per se, but to create alternative places where asylum could potentially be sought. These alternative asylum spaces concurred with the deinstitutionalisation process in placing individuals into the community but rebelled against its ideals of state intervention and standard practices of psychiatric care. The following section will detail Laing and Esterson's vision for a post-asylum landscape and showcase one such asylum project that came into being during this transitional period in mental health care.

The Philadelphia Association and Kingsley Hall

'Prayer For A Place
to dance and sing
to cry and scream
a place
to burst with agony and with ecstasy
a retreat
a place to go back to
        to go out from
a place
            to breathe again
a place to live
        to love
a place to be'[42]

Around the same time as Laing and Esterson were undertaking research into families at the Tavistock, they also drafted an application to the Foundation's Fund For Research in Psychiatry for resources to support a vision of 'a centre' in which the hypothesis that schizophrenia is 'much less of an organic illness … than a reaction to social stress of a particular kind' could be researched.[43] This space was imagined by Laing and Esterson to be 'a place of refuge for persons who have already disinte-

[42] R. D. Laing (1966) 'Prayer for a Place', Papers of R. D. Laing, Special Collections, University of Glasgow (MS Laing L221/22).

[43] R. D. Laing and A. Esterson (1964) 'Family and Schizophrenia Draft', Papers of R. D. Laing, Special Collections, University of Glasgow (MS Laing A627), n.p.

grated', and it would use methods of treatment such as milieu and family therapy in preference to the more physical treatments of insulin and electroconvulsive therapy.[44] The aim of this place, inspired by their previous family research and also by Laing's maturing vision of true asylum, was to allow the type of therapy to be matched to the personal and complex conditions experienced by the individual:

> In other words, unlike a mental hospital or similar institution, the social milieu will be adapted to the patient, and not the patient to the milieu.[45]

The Centre was arguably important not only because of the alternative therapies that it could offer, but more because of the particular approach to illnesses that it could take. Due to their awareness of different family dynamics, Laing and Esterson aimed to make this Centre a place which could provide 'a benign social matrix for persons with breakdowns to live in and work from, while taking part in individual and family therapeutic sessions'.[46] It was a model whose very conceptual foundations formed a larger project, the Philadelphia Association (PA) that attempted to bring their vision of true asylum to life.

The PA was founded officially in 1965 by R. D. Laing, Aaron Esterson, David Cooper, Clancy Sigal, Sidney Briskin, Raymond Blake and Joan Cunnold.[47] This diverse mix of individuals all believed in the need for a change in the way that mental illness was viewed and treated in society, in line with the broader shifts in mainstream policy driven by the 'care in the community' initiatives. The main aim of the PA was to set up household communities, hence 'homes' on the scale of the domestic homes previously researched, that could be 'places of sanctuary, asylum, refuge, dwelling' and also a vehicle to 'change the way the "facts" of "mental health" and "mental illness" are seen by many people'.[48] These household communities were designed to be dwellings 'where preconceptions are melted down in the direct experience of the wear and tear, agony

[44] Ibid., n.p.

[45] Ibid., n.p.

[46] Ibid., n.p.

[47] T. Itten (2005) 'All the Lonely People Where Do They All Come From: Facts, Feelings and Experience from the Philadelphia Association London', paper presented at the Association for Community Living Conference (New York).

[48] Philadelphia Association 'Leaflet', Papers of R. D. Laing, Special Collections, University of Glasgow (MS Laing L259/10).

and joy, excitement and boredom, hope and despair, of living together'.[49] It was this notion of living together in a particular environment that became pivotal in the project's approach to therapy, fuelling the notion that by fusing people together in one living environment the boundaries between patient and doctor/analyst could be broken down and healing could potentially take place. The PA was therefore an attempt to change the paradigm of 'doctor' and 'patient'/'us' and 'them', and therefore to engage with a new way of seeing the issues involved with mental health in the changing landscape of healthcare generally.

The Association set up a number of household communities throughout London, the most famous being Kingsley Hall: the others were The Archway Community, The Grove, Mayfield Road, and Portland Road.[50] When commenting on the character of the houses, Laing revealed, 'each place is a world unto itself, with its own drama'.[51] The distinctly different nature of each household was formed not only by the physical location and structure of the building itself, but also by the people who occupied these spaces. A former resident of the Archway Community remarked:

> The P.A. households, past and present, vary enormously in their openness to the outside world. Some are rather like a closed family system. Some are very cool to outsiders, seeing them as intruders, others have an open heart and are welcoming.[52]

The variety of responses to newcomers shows how emotionally tied the people living in these communities remained to their houses, as demonstrated by the recent collection of portraits from Kingsley Hall.[53] The lives and experiences that were lived out in the households of the PA are represented in stories that are told in very different ways. Many of these narratives centre on the daily experiences of Kingsley Hall and its members, but other stories offer important insights into the different worlds of the

[49] Ibid., n.p.

[50] Philadelphia Association (1978) 'Study Programme', Papers of R. D. Laing, Special Collections, University of Glasgow (MS Laing L259/2).

[51] Laing as quoted in Wykert, *Laing*, 28. The nature and location of many of these houses varied with many opening and closing in short periods of time. Although some of these houses, such as Kingsley Hall, had a very visible presence in the community, others wished to maintain a more understated existence.

[52] Itten, 'Lonely People', n.p.

[53] D. Harris (2012) *The Residents: Stories of Kingsley Hall, East London, 1965–1970 and the Experimental Community of R. D. Laing* (London: D. Harris).

Association and the many forms of 'therapy' taking place. As Laing himself noted, '[o]ne of the problems of recounting this story of Kingsley Hall is that the linear form of a verbal narrative is unadapted to the reality of the events, which are a set of patterns that undergo transformations, involving many people'.[54] Each dwelling's own individual character and structure became equally part of the project's strength and its weakness. Behind every door was supposedly a refuge supplying affection from others willing to accept another into a community, and this was the case for many. However, for others it was a place of bewilderment that felt as far away from sanctuary as the place that they had left behind. By delving deeper into the individual narratives of place in the following section, nuances appear about the significance of such alternative asylum spaces that are often lost and forgotten in the grand sweeping narratives surrounding deinstitutionalisation and community care.

'Asylum' Spaces and Post-Asylum Worlds

In returning to this chapter's opening quotation, layers of experience in these households begin to be revealed through investigating individual encounters with the place. Mike's memories of his time spent with the PA came as Kingsley Hall closed in the late 1960s. He remembers Kingsley Hall to be a place of physical and mental disarray:

> On the verge of becoming dilapidated and looking like a derelict warehouse, it [Kingsley Hall] was a far cry from the sanitary image of modern mental health. ... Most of the windows were broken and the heavy front doors, littered with obscene graffiti about 'nutters' and 'Meth drinkers', were battered and loose on their hinges.[55]

When the lease expired on Kingsley Hall a number of individuals, including Stuart and Mike, moved together into low-rent houses in an urban redevelopment area near Archway in North London, under the umbrella of the PA. While some members experienced an opportunity to live a so-called ordinary life with their mental health difficulties, others struggled to navigate through the darkened terrains of normal everyday existence.[56]

[54] Laing as quoted in Harris, *The Residents*, foreword.

[55] Mike, 'Memoir', 3.

[56] Harris, *The Residents*.

Mike recalls the intense difficulties of living in these homes with individuals experiencing severe mental health problems, including moments of violence and self-harm. An event described by Mike in relation to another member of the community, Peter, illustrates one of those situations:

> By the time I reached the downstairs hall I was no longer so anxious to know what was going on. Except for a thin bar of light shining beneath the common room door the house was dark. As I slowly began nudging the door open a blur of Peter slammed into it. I wedged my foot in and started to enter. ... Blood had spattered around the room and the walls and several record albums were smeared with it. ... At Kingsley Hall, I remembered, he had once beaten himself so badly that bruises were visible on his face and several days afterwards. I asked if I could help, hearing the lameness in my question even before I voiced it. 'No,' he said quietly. I felt embarrassed, as if I had intruded too deeply into his privacy.[57]

The difficulties documented by Mike's memoir bring to the fore the complex nature of many of these asylum spaces. For many living in these communities it was a last chance solution to long-term struggles and although the spaces themselves were designed to offer a 'secure place of refuge or shelter', the reality could be more fluid and unstable.[58] Mike notes that this is only one facet of his PA experience—although his memoir demonstrates some of the more painful aspects of living in such social contexts it also highlights that the value 'lies as much in what it strips away as in what it gives'.[59]

Further insights into the experience of these spaces of asylum was told through the medium of film and centred, once again, upon the Archway community in London. Called *Asylum* and directed in 1971 by film-maker Peter Robinson, this documentary is a voyage into the very heart of one of the PA households. The documentary, filmed by single cameraman Richard W. Adams, captures the role of observer from the outset as the camera follows the residents of the community as they go about their daily lives. The first half of the film is an uneasy time for the viewer, as they are deliberately thrown into confusion over who are the 'patients' and who are the 'doctors'. The bumpiness of the camera as it twists and turns upstairs and down, following these people and their activities, keeps the watcher

[57] Mike, 'Memoir', pp. 14–15.
[58] Laing, 'The What and Why of Asylum', n.p.
[59] Mike, 'Memoir', p. 29.

tuned in, desperate to figure out the strange happenings before their eyes. It is only when a father of one of the residents comes to pay for his son's stay and speaks to one of the residents that it becomes a little clearer who some of the patients may be, and it is interesting how 'crazy' many feel the father to be in his views on his son's condition. One reviewer noted:

> But I was most impressed by an outsider, a young male patient's business-like father, who, in paying his son's bills and announcing that he has hired a girl for the boy to date so as to build up his ego, suggests just enough about the world at large to make 'Asylum' seem a not unreasonable place to be.[60]

The film is introduced by Laing where he explains that 'asylum, I think, is to say, a safe place to be when you're frightened and alone', situating it in the heart of his small-scale vision for future psychiatric care.[61] The documentary has a disturbing but highly absorbing quality that attempts to shake the viewer vigorously out of their comfortable way of thinking about mental ill-health and its spaces of care. One reviewer found it impossible to walk away unaffected by this piece, exclaiming:

> Almost terrifyingly direct and involving—a model of *cinéma vérité*—Although the dialogue in ASYLUM is as bizarre as it is painful, truly tragicomic in a way that Beckett and Pinter would be stretched to equal, the characters can't be kept at an emotional distance. ASYLUM takes the key away from the audience—and makes excitingly dangerous connections.[62]

A further reviewer describes the film as 'artless and unpretentious, [*Asylum*] is really a communiqué from some other, unexplored subcontinent of the soul'.[63] The connections between the 'characters' and the audience that this film attempts to draw out highlight the roots of the PA project itself. The Association was designed on the premise that it would attempt to make the public question the traditional system of mental health 'care'. The notion that the 'patient' is someone who should be 'treated' in some capacity is something that the Association fought—and continues

[60] R. Greenspun (undated) 'The Screen: "Asylum", A Documentary', Papers of R. D. Laing, Special Collections, University of Glasgow (MS Laing T101), p. 19.

[61] Laing as quoted in ibid., p. 19.

[62] M. Knelman (1974) 'Review of Asylum, Toronto Globe and Mail', Papers of R. D. Laing, Special Collections, University of Glasgow (MS Laing T103).

[63] Wykert, 'Laing', pp. 16–17.

to fight—hard to overcome. For Laing and other fellow members, the understanding that 'therapy' has to be a fixed entity providing calculated 'solutions' was something they could not fully endorse; it was through developing the PA and its houses that they managed, in some small way, to allow people to see therapy in a variety of new lights and to continue to demonstrate that an individual experiencing mental health difficulties is a person requiring understanding, not a category necessitating explanation.

Conclusions: Seeking Asylum, Somewhere

> 'You need a frog to guide you in the bushes. So you won't get lost!' he declared with a burst of deep cackling laughter that trickled into stern, expectant silence.[64]

The experiments of Kingsley Hall, the Archway community and the entire project of the PA along with the legacy of Laing and Esterson's work on the family at the Tavistock Institute of Human Relations, showed that Laing and his colleagues were continuously reworking their own notions of asylum in the changing deinstitutional landscape. By paying attention to the inventing and implementing of small-scale spaces of care an alternative narrative of the deinstitutionalisation process in Britain is revealed that highlights the importance of individuals and small-scale strategies in the dawning of a new approach to psychiatric treatment in the period. Although many in the psychiatric profession were keen to dismiss these notions, some remained curious even if their own positions led them to take a more conventional approach. For example, Laing's supervisor at the Tavistock, Donald Winnicott remarked, after he had read *Sanity, Madness and the Family* and heard the many emerging stories from Kingsley Hall, that if he had his life all over again, he would have done what Laing was doing;[65] although the worlds of Laing and of psychiatry drifted further apart throughout the years, it was clear that he would continue to inspire questioning of the 'true' nature of the therapeutic encounter and the spaces of care it could produce.

Many of the examples used in this chapter to highlight the significance of place and spatial thinking in Laing and Esterson's imagining, and the designing and building of different types of asylum and post-asylum

[64] Mike, 'Memoir', p. 2.

[65] J. Clay (1996) *R. D. Laing: A Divided Self* (London: Hodder & Stoughton), p. 140.

spaces, have demonstrated the multitude of difficulties that are wrapped up in undertaking such ventures. The tragic circumstances of Mike and others' experiences of the Archway community deliberately highlight the pain, suffering and long-term trauma that experiencing mental health difficulties, at a range of scales, can have. In one sense this counters the often-quoted criticism of Laing's work that he romanticised mental suffering, but in another it shows the incredible difficulties involved in designing and implementing forms of asylum care. Asylum spaces in their multiple formats consistently reflect changing understandings of mental ill-health and by unpicking the foundations of some of these more experimental and alternative visions of asylum it is possible to add another dimension to the growing body of critical work on deinstitutionalisation and its aftermath.

In the wake of enormous political and legislative shifts, individuals such as Laing and Esterson dared to dream about other types of spaces where people like Peter could go to attempt to find a place of refuge and safety. While this may not have always been the case in reality, the inspiration behind such a drive comes from a desire to not see individuals such as Simon as patients to be treated but as people to be understood. It is this legacy that continues to haunt the remaining households of the PA and these small, often overlooked, alternative spaces of asylum have the potential to drive further imagining and developing of asylum spaces in years to come.

Acknowledgements Research for this chapter was made possible under the doctoral award funded by the Economic and Social Research Council (Award Number 030-2005-00165). Thanks to the editors of this collection for their insightful comments on earlier versions of this piece. Thanks also to Sarah Hepworth and the Special Collections at the University of Glasgow, for their help in locating the archival materials used in this paper, and to the Laing Estate for permission to access the R. D. Laing Collection. All R. D. Laing source material is copyright R. D. Laing Estate.

PART IV

After Deinstitutionalisation: Experiences, Challenges and Legacies

CHAPTER 11

The Basaglian Legacy in Italian Psychiatry: Remembering, Myth-Making and Crystallising

Elena Trivelli

Introduction

Let me begin with a rather sensational statement: in Italy, there are no psychiatric hospitals. Law 180, which abolished these institutions, was passed in 1978 and was largely the product of two decades of work by Franco Basaglia and his teams of physicians, mainly in Gorizia and Trieste in the north-east of the country. In this chapter, I focus on the latter, analysing Trieste's Mental Health Department as the representative centre of a 'Basaglian' legacy on a national and international scale.[1] As will become clear, the inverted commas are crucial. I interweave historical

[1] This legacy has attracted much international interest. See S. Ramon (1989) 'The Impact of the Italian Psychiatric Reforms on North American and British Professionals', *International Journal of Psychiatry*, 35, 120–27; N. Crossley (2006) Contesting Psychiatry: Social Movements in Mental Health (London and New York: Routledge).

E. Trivelli (✉)
Independent Scholar, Goldsmiths, University of London, London, UK
e-mail: e.trivelli@gold.ac.uk

D. Kritsotaki et al. (eds.), *Deinstitutionalisation and After*,
Mental Health in Historical Perspective,
DOI 10.1007/978-3-319-45360-6_11

accounts, media tropes and various critiques that constitute this legacy, framing them as enactments of the nostalgic feelings that have come to construe 'Basaglia' as a cultural currency and an idolised figure. I suggest that employing Basaglia in order to safeguard present 'models' of care significantly contrasts with the ideas that inspired his work.

In the first part of the chapter, I will mainly consider Basaglia's own writings, to analyse some salient aspects of his work in Gorizia, while also articulating the promulgation and implementation of Law 180, and its role in relation to the figure of Basaglia. In the second part, I will direct my attention to the aforementioned Trieste Mental Health Department, and to the tension between Basaglia and Basaglians. In discussing the symbols that might ossify a movement originally rooted in ideas of permanent crisis and self-questioning, I refer to the vicissitudes undergone by the sculpture 'Marco Cavallo' ('Marco the Horse') which has become a national symbol of deinstitutionalisation deeply tied to Basaglia's work in Trieste. My aim is not to evaluate the quality of healthcare services in the city, but to construct a platform for analysing the nostalgia that partly characterises the rhetoric and historiography around Italian deinstitutionalisation.

This chapter is based on an ethnography I conducted in Gorizia between 2011 and 2012, in which I similarly explored how Basaglia is remembered and discursively employed in constructing the city's identity and historical heritage.

Basaglia in Gorizia

Law 180 replaced a 1904 law that was the first to nationally regulate psychiatric practice across an Italy that had been unified only in 1861, and was marked by strong differences across its regions, most notably between the wealthier north and the impoverished south. Psychiatric hospitals were at that time supervised by the Interior Ministry and, from 1930, if the patient had not been discharged within 30 days his or her name was added to police criminal records.[2] It was only in 1968, with the passing of Law 431, that voluntary admission was introduced and the registration of patients into police criminal records abolished. Public asylums were at the time mostly populated by indigents. Hygiene in such institutions was extremely poor, and physical constraints, indiscriminate use of shock

[2] Law 36/1904, art. 8, and Penal Code, art. 604/1930. Notably, the latter did not apply to private institutions.

therapies and intrusive therapies were often the norm. Largely impermeable to psychoanalytic thinking, Italy had remained a 'distant spectator' to forms of alternative care that were developing, for instance, in the UK after the Second World War.[3] Strongly anchored to biomedical, organic, neurological and anatomical models of mental illness well into the 1950s, Italian psychiatry was generally practised in theoretical isolation and relative backwardness, since 'Lombrosianism' and social Darwinism maintained a strong hold on research in neurology.[4]

When Franco Basaglia (1924–1980) was appointed Director of the Psychiatric Hospital of Gorizia in 1961, he noted that, together with 'the enormous rooms, the shouting, the smell of urine, [and] the isolation chambers' there was a 'symbolic smell of shit'.[5] Influenced by Sartrean existentialism, Husserlian phenomenology, and by the work of Erving Goffman and Frantz Fanon, during his initial years in Gorizia, Basaglia developed a view of psychiatry as the rationalisation of suffering into an abstract category—the disease—that worked as an ideological 'double' over the material distress of the subject.[6] The symbolic stench of the hospital was produced by the 'institutional regression' that psychiatric authority had legitimated in such a space—a concept akin to Russell Burton's 'institutional neurosis'—whereby the asylum had become not a place of care, but one of mere segregation which had exacerbated patients' distress.[7]

From 1962, Basaglia began to gather a team of young physicians in Gorizia. They soon started to abolish physical constraints and shock therapies, gradually opening the wards and allowing patients to leave the

[3] M. Donnelly (1992) *The Politics of Mental Health in Italy* (London and New York: Routledge), p. 25.

[4] A. Tagliavini (1985) 'Aspects of the History of Psychiatry in Italy in the Second Half of the Nineteenth Century' in W. F. Bynum, R. Porter and M. Sheperd (eds) *The Anatomy of Madness: Essays in the History of Psychiatry* (London and New York: Tavistock), pp. 175–96.

[5] E. Venturini (2010) 'L' Incidente di Imola' in E. Venturini, D. Casagrande and L. Toresini (eds) *Il Folle Reato: Il Rapporto tra la Responsabilità dello Psichiatra e la Imputabilità del Paziente* (Milan: FrancoAngeli), pp. 28–103, 57; F. Basaglia (2000) *Conferenze Brasiliane* (Milan: Raffaello Cortina), p. 49.

[6] F. Basaglia (1973) 'Che cos'è la Psichiatria?' in F. Basaglia (ed.) *Che cos'è la Psichiatria?* (Turin: Einaudi [1967]), pp. 3–13; F. Basaglia and F. Ongaro Basaglia (1971) 'La Malattia e il Suo Doppio' in F. Basaglia and F. Ongaro Basaglia (eds) *La Maggioranza Deviante: L' Ideologia del Controllo Sociale Totale* (Turin: Einaudi), pp. 133–39.

[7] F. Basaglia (1973) 'La Libertà Comunitaria come Alternativa alla Regressione Istituzionale' in Basaglia (ed.) *Che cos'è la Psichiatria?*, pp. 15–31; R. Burton (1976) *Institutional Neurosis* (Bristol: John Wright & Sons [1959]).

premises accompanied by nurses. Initially inspired by British models of Therapeutic Communities (TCs), assemblies began to take place inside the psychiatric hospital. However, unlike the experiences of Maxwell Jones or David Cooper in Britain, these were not aimed at group psychotherapy.[8] Rather, it was the process of socialisation and the opportunity to discuss and engage with the management and the politics of the hospital that were deemed therapeutic. In other words, it was the process of organising the TC, and not the TC itself, that was beneficial.[9] While not conceived as purely a product or a consequence of social conflicts, mental distress was understood to arise within the realm of the social, and it was in this same realm that it should be addressed. The need for the patient to return to society implied a radical interrogation of the nature of the segregating institution, and a process of questioning not what mental illness was, but rather what psychiatry was, and what it had so far achieved towards alleviating distress.[10]

The team in Gorizia soon began to radicalise their ideas, indicating that an optimal TC would resemble a 'golden cage': a place of humane treatment, but still a place of segregation. As in previous forms of extra-asylum care, the patient might be physically liberated, only to be then 'astutely tamed', in a fashion not dissimilar to the '"liberation" brought about by Pinel and Tuke'.[11] The analysis of the 'humanitarian' components of the TC implied understanding it not as a solution, but as a transitory phase. As Basaglia put it, 'if the Therapeutic Community is accepted as a model, the process of transformation is halted, and reduced to forms of adjustment'.[12] I particularly want to stress that Basaglia's work in alterna-

[8] D. Cooper (1967) *Psychiatry and Anti-Psychiatry* (London: Tavistock); M. Jones (1952) *Social Psychiatry: A Study of Therapeutic Communities* (London: Tavistock).

[9] F. Basaglia (1981) 'La "Comunità Terapeutica" come Base di un Servizio Psichiatrico: Realtà e Prospettive' in F. Ongaro Basaglia (ed.) *Basaglia: Scritti I: 1953–1968: Dalla Psichiatria Fenomenologica all'Esperienza di Gorizia* (Turin: Einaudi), pp. 259–82; F. Ongaro Basaglia (1971) 'Rovesciamento Istituzionale e Finalità Comune' in F. Basaglia (ed.) *L'Istituzione Negata: Rapporto da un Ospedale Psichiatrico* (Turin: Einaudi [1968]), pp. 321–35.

[10] F. Basaglia (1971) 'Le Istituzioni della Violenza' in Basaglia (ed.) *L'Istituzione Negata*, 113–51; Basaglia (1973), 'Che Cos'è la Psichiatria?'.

[11] G. Jervis and L. Schittar (1973) 'Storia e Politica in Psichiatria: Alcune Proposte di Studio' in Basaglia (ed.) *Che Cos'è la Psichiatria?*, pp. 171–202, 178; M. Foucault (2009) *History of Madness* (London: Routledge [1972]), p. 107.

[12] F. Basaglia (1982) 'Le Istituzioni della Violenza e le Istituzioni della Tolleranza' in F. Ongaro Basaglia (ed.) *Scritti II: 1968–1980: Dall'Apertura del Manicomio alla Nuova*

tive psychiatry was founded upon a continuous questioning of power relationships. It entailed constant dialectic negotiations and a critical suspicion of therapeutic 'models', which hold the intrinsic risk of ossifying more subtle forms of power into disciplinary apparatuses for the organisation of madness.[13]

The work in Gorizia was thus better conceived of as an 'anti-model', since its constant self-questioning implied a scepticism towards any exemplary standards. As Félix Guattari put it, such work was conducted with a 'permanent refusal of all self-satisfaction'.[14] In this sense, 'the Therapeutic Community is a form of freedom while on the make ... But it would become a form of oppression, if one was to enclose it in a fixed schema.'[15] The creation of a 'good institutional product' would therefore not only fail to 'produce a crisis in daily asylum practice' and overturn the system of oppression that sustains the institution, but it would also be reabsorbed into a reformist logic, 'incorporated as an ideology, a new label'.[16]

From Gorizia to Trieste and to Law 180

At this point, the work carried out in Gorizia needed to physically exit the asylum, and the team pushed for the constitution of external mental health centres, which would in time become the alternative to the monolithic hospital, rather than supplementing it. This project, however, led to clashes with the local administration in Gorizia. Basaglia chose to leave Gorizia in 1968, and in 1971 he obtained the position of Director of the San Giovanni Psychiatric Hospital in the nearby city of Trieste.

Here, Basaglia and a new team of physicians immediately coupled the Therapeutic Community with the opening of external centres. They held a number of music festivals and cultural events inside the hospital, also

Legge sull'Assistenza Psichiatrica (Turin: Einaudi), pp. 80–86, 84.

[13] In line with Michel Foucault's critique of psychiatry as a system of disciplinary apparatuses, the work in Gorizia can be seen as a practical interrogation of psychiatric power and a practical questioning of its mechanisms within the asylum. See M. Foucault (2008) *Psychiatric Power: Lectures at the Collège de France 1973–1974* (Basingstoke and New York: Palgrave MacMillan [2003]).

[14] F. Guattari (2009) 'Guerrilla in Psychiatry: Franco Basaglia' in S. Lotringer (ed.) *Chaosophy: Texts and Interviews 1972–1977* (Los Angeles: Semiotext(e)), pp. 119–23, 121.

[15] F. Basaglia (1982) 'Appunti di Psichiatria Istituzionale' in Ongaro Basaglia (ed.) *Scritti II: 1968–1980*, pp. 47–72, 65.

[16] Basaglia, 'Le Istituzioni della Violenza', p. 149; F. Basaglia (1982) 'Introduzione a Lo Psicanalismo' in Ongaro Basaglia (ed.) *Scritti II: 1968–1980*, pp. 349–53, 351.

hosting performances by artists such as Ornette Coleman and Dario Fo, which residents of the city began to attend. From 1973, social cooperatives began to offer employment opportunities for patients, who were gradually discharged through the implementation of group housing, or continued to live in the hospital as voluntary 'guests'. In 1973, Basaglia contributed to the foundation of the pressure group Democratic Psychiatry, which played an important role in the drafting and passing of Law 180 in 1978. In January 1977, he declared that the hospital was closed, and its formal closure was announced in April 1980, only a few months before Basaglia's sudden death from a brain tumour.[17]

Law 180 was therefore a product of almost two decades of work in Gorizia and Trieste. It was also the outcome of some specific historical, political and geographical contingencies which deeply influenced its modes and levels of implementation. It was promulgated during a season of intense public participation in political decision-making and liberal reforms of the Civil Code, such as the 1974 referendum on divorce and 1981 referendum on voluntary pregnancy termination. These were also the 'Years of Lead' (1972–1982), when the rise of autonomist movements from the late 1960s was coupled with a series of terrorist attacks by extremist political fringes and strong security measures across the country, which were part of a so-called 'strategy of tension'.[18] While left-wing parties were hegemonic in regional and municipal constituencies, the moderate conservative party Christian Democracy maintained a parliamentary majority.

Law 180 passed stealthily on 13 May 1978, four days after the discovery of the body of Aldo Moro, the leader of Christian Democracy, who had been assassinated by the Red Brigades, a well-known terrorist cell on the fringes of the extreme left. The law that reformed the national psychi-

[17] P. Dell'Acqua (2007) *Non ho l'Arma che Uccide il Leone* (Viterbo: Stampa Alternativa [1980]).

[18] A. Cento Bull (2007) *Italian Neofascism: The Strategy of Tension and the Politics of Nonreconciliation* (New York: Berghahn Books). In recent years, it has been suggested that many of these terrorist attacks were manoeuvred by the Italian state, in collaboration with the CIA, in order to encourage a right-wing shift in the electorate. See R. Bartali (2006) 'The Red Brigades and the Moro Kidnapping: Secrets and Lies' in A. Cento Bull and A. Giorgio (eds) *Speaking Out and Silencing: Culture, Society and Politics in Italy in the 1970s* (London: Money Publishing), pp. 140–60; P.Willan (2001) 'Terrorists "Helped by C.I.A." to Stop the Rise of the Left in Italy', *The Guardian*, 26 March.

atric system was thus quickly passed, within a stalemated political system and a stagnant economic situation.[19]

Law 180 was incorporated with Law 833 in December 1978. The latter instituted the National Health Service, and created Local Health Units for specific catchment areas—with Mental Health Departments as a subcategory—which would be managed by individual regions. Law 833 allowed regional-level administrations to oversee the implementation of Law 180, providing no specific guidance or state supervision. The degree of implementation depended on the budget that each region allocated to its mental health services. The reform was therefore applied heterogeneously, like 'spots on the coat of a leopard', with particularly marked differences between the north and the south, and several psychiatric hospitals maintained their functions after 1978.[20] Regrettably, any attempt to quantitatively evaluate the implementation is hampered by a lack of consistent data collected across the decades.[21]

In 1994, the first Mental Health Plan was issued by the Italian Parliament. Law 724 (1994) and Law 662 (1996) imposed the closure of the remaining psychiatric hospitals by December 1996, introducing sanctions for defaulting regions. Law 449 (1997) postponed the deadline to December 1998, with a second Mental Health Plan. However, psychiatric practice in Italy remains marked by significant regional differences, most notably in the distribution of public and private services in the north and the south.[22] Private institutions are not entirely subjected to the impositions of

[19] In 1977, a coalition between the Communist Party and the Radical Party began to collect signatures, in order to reach a referendum that would abrogate Law 36 (1904). While the minimum quota to set in motion the mechanism for a national referendum is 500,000 signatures, the campaign for the abrogation of Law 36 quickly collected over 700,000. To avoid the risk of a referendum—which, in Italy, can only be abrogative—and a consequential legal vacuum, the government quickly produced the draft of Law 180. See M. Donnelly, *The Politics of Mental Health in Italy*; P. Ginsborg (1990) *A History of Contemporary Italy* (London: Penguin).

[20] V. Pastore (1994) 'Vita e Miracoli della Legge 180' in V. Fiorino (ed.) *"Rivoltare il Mondo, Abolire la Miseria": Un Itinerario Dentro l'Utopia di Franco Basaglia, 1953–1980* (Pisa: ETS), pp. 96–115, 109; G. B. Palermo (1991) 'The 1978 Italian Mental Health Law—A Personal Evaluation: A Review', *Journal of the Royal Society of Medicine*, 84, 99–102.

[21] L. Burti and P. Benson (1996) 'Psychiatric Reform in Italy: Developments since 1978', *International Journal of Law and Psychiatry*, 19, 373–90.

[22] F. Sironi (2013) 'La Legge Basaglia Funziona o No?', *L'Espresso*, March, at http://espresso.repubblica.it, accessed 4 October 2013.

Law 180, and in 2002 they supplied 43 per cent of inpatient psychiatric beds, often outnumbering public services in the south.[23]

Initial reactions to Law 180 often denounced the 'anti-social consequences of a radical law', which allegedly disdained the use of pharmaceuticals and discarded biological views in favour of sociopolitical aetiologies.[24] While it is mainly renowned for 'closing psychiatric hospitals' in Italy, Law 180 is primarily concerned with establishing that the norm for treatment is voluntary. It regulates the dynamics of compulsory admissions and it stresses the importance of local outpatient clinics and mental health centres.[25]

A number of family associations began to form in the early 1980s and many of these are still actively involved in lobbying for the hardening of the law, which they define as 'inefficiency, cruelty, exploitation, and superstition'.[26] Even in Trieste, in the 1980s, it was being described as 'a terrible experiment on human beings'.[27] Since its promulgation, Law

[23] A. Lora (2009) 'An Overview of the Mental Health System in Italy', *Annali Istituto Superiore della Sanità*, 45:1, 5–16; ISTAT (2006), L'Ospedalizzazione di Pazienti Affetti da Disturbi Psichici 1999–2002, n.11, at http://www. istat.it, accessed 15 July 2013. See also A. Signorelli (2009) 'Profonda/mente Sud: La 180 si è Fermata ad Eboli' in P. Lupattelli (ed.) *I Basagliati: Percorsi di Libertà* (Perugia: CRACE), pp. 157–62; G. De Girolamo and M. Cozza (2000) 'The Italian Psychiatric Reform: A 20-Year Perspective', *International Journal of Law and Psychiatry*, 23, 197–214.

[24] R. Papeschi (1985) 'The Denial of the Institution: A Critical Review of Franco Basaglia's Writing', *British Journal of Psychiatry*, 146, 247–54, 253.

[25] Law 180 regulates compulsory admission to a maximum of seven days, extendible week by week, and it decrees that these procedures are to be undersigned by two practitioners, as well as the mayor of the patient's municipality. It rules that psychiatric wards should be included as small structures within or in the proximity of general hospitals, with a maximum of 15 inpatients beds. Crucially, it has involved the rethinking of the role of clinical staff in psychiatric care. This differs, for example, from the British deinstitutionalisation process that began in the 1990s as a managerial reconfiguration of services. See J. Jones (2000) 'Mental Health Care Reforms in Britain and Italy since 1950: A Cross-National Comparative Study', *Health and Place*, 6:3, 171–87.

[26] http://vittimedella180.org. See also http://arap.it, accessed 4 October 2013. The primary criticisms that still underpin the campaigns of many of these family associations revolve around the difficulties in accessing inpatient structures for non acute cases, as well as the urgency of more prolonged periods of stay for acute patients. See G. Bersani (2009) 'The Italian Law "180" Thirty Years After: From the Unmet Needs of Sick People and Families the Duty of a Critical Reflection', *Annali Istituto Superiore della Sanità*, 45:1, 27–32.

[27] U. Dinelli (1987) 'La Legge 180 Deve Cambiare', *Il Piccolo*, 16 September, p. 7.

180 has also been widely criticised by members of the Italian right, and a number of proposals for its revision have been presented over the years.[28]

'The Basaglia Law'

Law 180 is often popularly referred to as 'the Basaglia Law'. This is not only technically incorrect, since the law was drafted by member of parliament Bruno Orsini, but it also has important implications for construing and maintaining a 'Basaglia legacy'. In her history of Italian psychiatry, Valeria Babini notes that 'Law 180 was not "the Basaglia law" .... It should not even have been a law in itself, but part of the National Health System. And yet, that encapsulation would separate and distinguish it, making it an easier target for attacks.'[29] Crucially, the Basaglia Law embodies the myth of Franco Basaglia, 'a man who managed to transform his extraordinary sensibility into intelligence, intelligence into theory, and theory into a concrete endeavour'.[30] By 1973, he was already portrayed as 'the Messiah', 'the Sun King of antipsychiatry', the 'anti-Pope, pontiff of psychiatric dissent in Italy' and as Babini puts it, 'Italy, in the 1970s, was "Basaglia's country".'[31]

To analyse the history of Italian psychiatry means to consider also how this rhetoric might shape present modes of engagement with such history. In fact, the Basaglians who took part in this season of radical changes tend to respond to the attacks on Law 180 by 'trenching into the fights of the 1960s and 1970s'.[32] These debates are 'still crystallised on the division between psychiatry and anti-psychiatry, liberators and oppressors, biological or social approach', whereby Basaglia is invoked as an idolised figure of such battles.[33]

[28] L. Benevelli (2010) 'Le Schede e i Commenti agli Ultimi 9 Disegni di Legge di Modifica', *Forum Salute Mentale*, 6 June, at http://news-forumsalutementale.it, accessed 27 July 2013.

[29] V. Babini (2009) *Liberi Tutti: Manicomi e Psichiatri in Italia: Una Storia del Novecento* (Bologna: Il Mulino), p. 290.

[30] M. Tommasini (1994) 'Le Cooperative della 180 a Parma: Una Realtà Possibile' in Fiorino (ed.) *"Rivoltare il Mondo, Abolire la Miseria"*, pp. 115–26, 116.

[31] F. Chiocci (1973) 'Sfida anche il Codice Penale il Profeta dell'Antipsichiatria', *Il Tempo*, 15 July, p. 24; G. Zincone (1973) 'Il Pioniere dei Manicomi "Aperti"', *Il Corriere della Sera*, 4 January, p. 16; Babini, *Liberi Tutti*, p. 177.

[32] M. Fiorani (2010) *Bibliografia di Storia della Psichiatria Italiana 1991–2010* (Florence: Florence University Press), pp. 17–18.

[33] Ibid., p. 13.

Slightly trite and edulcorated accounts are recurrent tropes in periodical revivals of his work. For instance, while the 2000 film *La Seconda Ombra* (*The Second Shadow*) has been criticised by some as a 'sanctification' of his figure, in his recent biography on Franco Basaglia, Oreste Pivetta defines it as a 'biographical account'.[34] Pivetta also largely skates over the divergences inside the Gorizia team, which have been detailed by Giovanni Jervis, a member of the original group.[35] Over the years, many Basaglians have accused Jervis of 'historical revisionism' for reporting the dissent in the team and Basaglia's authoritarian character as 'aspects that the public did not want to hear'.[36]

The institutionalisation of his charisma made Basaglia almost the sole representative of a cultural movement that, in fact, involved a variety of actors, such as physicians and nurses who challenged traditional methods of care, as well as patients who at times took leading roles in Gorizia and Trieste. Caught between the myth and its abuse, his name has the power to legitimate, credit or reproach. Doing things 'in a Basaglian method', 'in a Basaglian spirit', 'as Basaglia would have done', is an insistent leitmotif of this legacy, as a recent publication that coins the term 'the Basagliated' exemplifies.[37] Basaglia has been employed frequently as a currency that is mainly associated with deinstitutionalisation practices in Trieste, after it became 'the new Gorizia'.[38] In fact, there is occasional conflation in media accounts, for 'the two cities are close, and they get confused as if they were the same thing', with 'the effect that elsewhere one hears about the experience of Trieste as if it has been the first and the most important'.[39]

[34] S. Agosti (2000) *La Seconda Ombra* (Italy: RAI) (feature film); G. Pirella (2000) 'Basaglia "Santo Taumaturgo"', *Isonzo Soča*, 35, 5; O. Pivetta (2012) *Franco Basaglia: Il Dottore dei Matti: La Biografia* (Milan: Dalai Editore), p. 124.

[35] G. Jervis (1977) *Il Buon Rieducatore: Scritti sugli Usi della Psichiatria e della Psicoanalisi* (Milan: Feltrinelli); G. Jervis and G. Corbellini (2008) *La Razionalità Negata: Psichiatria e Antipsichiatria in Italia* (Turin: Bollati Boringhieri).

[36] Jervis, *Il Buon Rieducatore*, p. 25; Jervis and Corbellini, *La Razionalità Negata*; E. Venturini (2008) 'Da Gorizia Replica a Jervis: Basaglia era Libero dai Poteri', *Il Piccolo*, 31 October, p. 9.

[37] Lupattelli, *I Basagliati*.

[38] R. Lugli (1974) 'Lager per 10000 Alienati', *La Stampa*, 24 June, p. 13.

[39] S. Simoncini (1996) 'L' Istituzione Dimenticata: L'Esperienza Psichiatrica a Gorizia negli Anni Sessanta e Settanta—Intervista ad Agostino Pirella', *Isonzo Soča*, 21, 10–14, 12.

Deinstitutionalisation in Trieste

Since 1974, the World Health Organization has conducted extensive research on Trieste as a model for deinstitutionalisation. In the midst of 'the highly publicized failures of "deinstitutionalization" in major American cities', Trieste proved that 'there [was] at least one city in the world that seem[ed] to have made it work'.[40] After his appointment as Hospital Director in 1971, Basaglia relied strategically on media visibility to attract national and international interest.[41] While Gorizia had also been visited by volunteers, practitioners and journalists, the work in Trieste was rendered spectacular through events such as arranging a flight with 100 patients in 1975, in which David Cooper (prominent in the anti-psychiatry movement) took part, or hosting the International Réseau of Alternative Psychiatry in 1977. As Babini puts it, Trieste was becoming what 'Freud's Vienna' once was.[42] In 1987, the Trieste Mental Health Department officially became a WHO Collaborating Centre, and it has since promoted a programme of conferences, internships, summer schools and international research groups, also collaborating with University College London, and with non-governmental organisation (NGO) groups such as the International Mental Health Collaborating Network. As a city with 'lessons to offer', in 2005 the Department was elected Leading Collaborating Centre for Service at the Helsinki Declaration and Mental Health Action Plan.[43] In 2010 it was once again nominated a WHO Collaborating Centre for Research and Training in Mental Health, and it hosted the international meeting 'Trieste 2010: What is "Mental Health"? Towards a Global Network of Community Health', organised by the NGO association Permanent Conference for Mental Health Worldwide.

While several psychiatric hospitals in the country operated a successful policy of deinstitutionalisation, it has been reported that a 'Basaglian cure'

[40] World Health Organization (1978) *Changing Patterns in Mental Health Care: Report on a WHO Working Group* (Cologne: Regional Office for Europe), p. 9; P. M. Boffey (1984) 'Treating the Mentally Ill: Trieste's Lesson', *New York Times*, 17 January, at http://nytimes.com, accessed 27 July 2013.

[41] N. Pitrelli (2004) *L'Uomo che Restituì la Parola ai Matti: Franco Basaglia, la Comunicazione e la Fine dei Manicomi* (Rome: Editori Riuniti).

[42] Babini, *Liberi Tutti*, p. 269.

[43] A. Cohen and B. Saraceno (2002) 'The Risk of Freedom: Mental Health Services in Trieste' in A. Cohen and B. Saraceno (eds) *World Mental Health Casebook: Social and Mental Health Programs in Low-Income Countries* (New York: Kluwer Academic/Plenum Publishers), pp. 191–220, 191.

has found 'ample application' 'only in Trieste'.[44] Here, Basaglia is heavily employed for the image of its Mental Health Department—most strikingly, on its website—which 'has represented and still represents a place of efficiency and innovative projects that is very peculiar, if not unique'.[45] In fact, Trieste has consistently produced cultural and medical programmes and syllabi, often focusing on Basaglia's work in the city as an important element of its local heritage, and it presents a model of care that some hope 'will spread across Italy'.[46] Noting these tropes is crucial for interpreting the description of the latest proposal of reform of Law 180 as 'an attack against Trieste', which has been insistently defined as the national 'small fort of mental health' and 'the only bastion against a movement that tends to reaffirm the asylum model'.[47] In fact, the city has been recently described as 'the symbol of law 180'.[48]

Notwithstanding such distinguished descriptions, some less widely broadcasted but important issues beleaguer the city. I will now turn my attention to the Mental Health Department in Trieste, outlining some of the allegations of practices of non-communication and authoritarianism, as well as internal controversies between Basaglians. My aim is neither to verify the accuracy of such charges nor to propose—or reproduce—dichotomies or binaries such as 'for' or 'against Trieste'. Rather, I intend to bring these often unacknowledged issues to attention, and to read the Department's employment of Basaglia—the idol and the currency—in relation to the foundational stances of his work, as outlined above.

[44] G. Ziani (2009) 'Lavoro Contro il Disagio Mentale: Trieste Capofila in Italia', *Il Piccolo*, 16 January, p. 20.

[45] See http://triestesalutementale.it, accessed 4 October 2013; B. Norcio (2009) 'La Legge 180, il Sistema di Salute Mentale in Italia, l' Esperienza Pilota di Trieste: É Necessario un Cambiamento?' in Lupattelli (ed.) *I Basagliati*, pp. 170–77, 174.

[46] Norcio, 'La Legge 180', p. 176; See, for example, Permanent Conference for Mental Health Franco Basaglia Worldwide 'Trieste 2010: what is "Mental Health?"' (February 2010), 'Beyond the Walls' (April 2011), or the 'Franca e Franco Basaglia Summer School', at http://triestesalutementale.it, accessed 4 October 2013.

[47] R. Cosolini (2009) 'Pd: Legge 180, un Attacco a Trieste', *Il Piccolo*, 1 February, at http://ilpiccolo.geolocal.it, accessed 3 October 2013; G. Ziani (2013) 'Signorelli (Dsm): Ho i Titoli. Attaccata perché Donna', *Il Piccolo*, 9 February, at http://ilpiccolo.geolocal.it, accessed 3 October 2013; R. D'Alessandro (2008) *Lo Specchio Rimosso: Individuo, Società, Follia da Goffman a Basaglia* (Milan: FrancoAngeli), p. 261.

[48] A. Cassin (2009) 'Operazione Ignobile', *Il Piccolo*, 3 February, at http://ilpiccolo.geolocal.it, accessed 20 September 2013.

The Department has been repeatedly accused of 'non-dialogue' by several family associations.[49] The local bureau of the national family association ARAP (Associazione per la Riforma dell'Assistenza Psichiatrica) denounces 'deafness, misunderstanding and disinterest', as well as 'walls of silence' when relating to the Mental Health Department. ARAP, which was born in 1982 with the precise aim of modifying Law 180, accuses the Department of a 'propaganda' which 'offers a deformed and idyllic overview of Italian psychiatry'.[50] Even more bluntly, a local independent family association describes the management as a 'totalitarian regime', based on a 'social nepotistic system' managed by the 'hierarchs of a pseudo-Basaglian theocracy'.[51] In a similar vein, it has been accused of 'pursuing business interests and defending personal privileges', where a 'fanatical and nepotistic management of power is implemented through career obstructionism and a dogmatic planning of information'.[52] Often, such accusations are also entangled with explicit forms of political bias, where the spectre of communism and Marxism are often rhetorically mobilised in tendentious descriptions of the Department's work, coupled with allegations of anti-scientific and sociogenetic approaches to mental distress.

Many former and current nurses from Gorizia currently speak of Trieste as an 'ivory tower'. The bitterness implicit in this expression refers on the one hand to the imminent threat that the six regional local health units might be soon unified into a single institution, coordinated by Trieste, under a plan significantly referred to as the 'Vast Area Project'. This would entail a considerable loss of autonomy in the management of local resources for other health units in the region. On the other hand, the malice suggested by the rhetoric of an 'ivory tower' alludes to practices of remembering Basaglia's work in the two cities. As two former nurses from Gorizia declared to the regional newspaper *Il Piccolo*, when trying to organise commemorative events around Basaglia's death or the anniversary of Law 180, Gorizia is often 'robbed' of these initiatives by Trieste.[53]

[49] C. Lodi (2009) 'Dell'Acqua Denunciato di Non Dialogo', *Libero*, 14 January, at http://liberoquotidiano.it, accessed 3 October 2013.

[50] http://sospsiche.it/associazioni-aderenti-al-sito/aderenti-al-sito/arap-trieste/, accessed 3 October 2013.

[51] http://giuliocomuzzi.it, accessed 4 October 2013.

[52] A. Segatori (2010) *Oltre l'Utopia Basagliana: Per un Nuovo Paradigma della Psichiatria: Fondamenti Psicosociologici e Sociopolitici della Legge 180* (Milan and Udine: Mimesis), pp. 24, 34.

[53] F. Fain (2010) 'Gorizia Rimpiange Basaglia. "Ora Almeno una Via"', *Il Piccolo*, 13 February, at http://ilpiccolo.geolocal.it, accessed 27 July 2013.

Basaglia and Basaglians

Simultaneously presented as a model of good practice, the emblem of Basaglia's accomplishment and the repository of his legacy, Trieste is also the place where, according to Jervis, Basaglia became surrounded by a group of practitioners that were 'more Basaglian than himself' and where, Jervis continues, he became a victim of his own fame.[54] Local psychiatrist Adriano Segatori highlights the importance of distinguishing between Basaglia and Basaglians. He fiercely defines the latter as a 'church' or a 'sect', and this legacy has also been framed as one of 'inane and impromptu devotees'.[55] Even among these 'devotees', however, 'being a Basaglian' appears to be a highly contested designation, whose meaning is profoundly controversial.

In fact, a degree of commotion has animated the nomination of a new Department Director in 2013, with the retirement of Peppe dell'Acqua, who had been a direct collaborator of Basaglia in Trieste in the 1970s. The 'rumpus' arose when the outgoing Director commented on his Facebook page, shortly after the new Director had been nominated by the local health unit, that 'in the course of these years, an incompetent and strategic management, petty and compliant, has occupied the health unit', with the aim of 'erasing Basaglia and the communist party. They have unleashed boycotters, informers, stupid idiots.'[56] Notably, both parties involved in this altercation—the outgoing Director Dell'Acqua and the newly appointed Assunta Signorelli—are former collaborators of Franco Basaglia, self-identified and self-proclaimed Basaglians. However, as local journalist Fabio Dorigo notes, such internal conflicts express the supposedly divergent currents of 'orthodox Basaglians (pure and hard)' and 'heretic Basaglians (collaborationists)'. As Dorigo continues, Basaglia perhaps 'could not imagine that the showdown of the psychiatry that closed the asylums would take place around his name and in the Trieste Mental Health Department', nationally and internationally renowned as 'the repository of his revolution'.[57] However, it is precisely 'around his name', his mythicised figure, his charisma and accomplishments that, as Segatori

[54] Jervis and Corbellini, *La Razionalità Negata*, p. 94.

[55] Segatori, *Oltre l'Utopia Basagliana*, p. 16; C. Magris (2009) 'Così è stato Tradito Basaglia: Successi e Lacune di una Legge', *Il Corriere della Sera*, 10 May, 34–35, 35.

[56] F. Dorigo (2013) 'Dsm, Cambio al Vertice da Mazzina a Signorelli. E Dell'Acqua Attacca', *Il Piccolo*, 8 February, at http://ilpiccolo.geolocal.it, accessed 20 September 2013.

[57] Dorigo, 'Dsm, Cambio al Vertice da Mezzina a Signorelli'.

puts it, 'Basaglianism' has turned into a 'Gospel' and a 'totem', or an 'unquestionable, untouchable and timeless doctrinaire system'.[58]

Historian Matteo Fiorani frames this 'timelessness' as a 'crystallisation' and a 'selective attention for the past', where affectionate memories prevail over historical chronicles in many of the accounts produced by those who directly experienced this season of changes in the 1960s and 1970s.[59] Historiographical work on Italian psychiatry indeed tends to draw a linear and progressive narrative, generally portraying the 1960s and 1970s as the peak of such progress.[60] Many accounts are often tied to the imprinting that this season of social changes gave to the history of Italian psychiatry, and they tend to present a 'crystallised conflict' between 'good' and 'bad' practice, 'Basaglians' and 'anti-Basaglians'.[61] In this context, Jervis pungently notes that

> if a foreigner, or a historian in the future, wished to get a clear idea of what the 'new psychiatry' in Italy has been, it would be extremely useful to read a few personal, honest and sincere accounts, with the lights and their shadows, the hopes, the illusions and the disappointments: it would be much less useful to read the official histories.[62]

Nostalgia

Nostalgia is a useful framework for analysing processes of myth-making and currency-making, as well as practices of crystallisation and selective attention in the historiography of Italian psychiatry. I here refer to nostalgia not as 'the inner space of an individual psyche', but as the 'intermediary between collective and individual memory', a mediation between the past and the present, and a socialised affective process of meaning making.[63] In this respect, Svetlana Boym frames the encounter between memory and history as a 'romance with the past', which calls for 'the

[58] Segatori, *Oltre l' Utopia Basagliana*, pp. 318, 14, 34.

[59] A. Fiorani (2010), *Bibliografia di Storia della Psichiatria Italiana 1991–2010*, pp. 31, 17. See Dell'Acqua, *Non ho l'Arma che Uccide il Leone.*

[60] P. Guarnieri (1991) *La Storia della Psichiatria: Un Secolo di Studi in Italia* (Florence: Leo S. Olschki).

[61] Fiorani, *Bibliografia di Storia della Psichiatria Italiana.*

[62] G. Jervis (1980) 'Intervento' in L. Onnis and G. Lo Russo (eds) *Dove Va la Psichiatria? Pareri a Confronto su Salute Mentale e Manicomi in Italia dopo la Nuova Legge* (Milan: Feltrinelli), pp. 68–97, 75.

[63] S. Boym (2001) *The Future of Nostalgia* (New York: Basic Books), pp. 41, 54.

repetition of the unrepeatable, the materialisation of the immaterial'.[64] At the intersections between the private and the public, the present and the past, the individual and the social, nostalgic feelings tangle and confound the past and the present, as well as real and imaginary events, affecting the remembering of the past, the reimagining of the present, and the envisioning of the future. In this sense, 'history' becomes not a rigid element, but rather something constituted by a set of present practices that are predicated upon the unpredictability and potential rewriting of the past. Following Boym's analysis, I maintain that, while it can be deployed for political purposes, nostalgia is, above all, a form of anguish, or an 'ache of temporal distance and displacement'.[65] Examining the etymology of the term, Boym distinguishes between forms of restorative nostalgia—which stress nostos, as the return or the will to rebuild—and reflective nostalgia—which emphasise algos, or the pain of loss.

I suggest that, implicit in the employment of Basaglia as a cultural, historical and political currency in Trieste, is an oscillation between forms of restorative and reflective nostalgia, which employ Basaglia's work both as a 'value for the present', or a 'perfect snapshot', and also cherish 'shattered fragments of memory'.[66] Between algos and nostos there are the images, projections and defensive mechanisms one might resort to when filling in the gaps of what one does not know—or what one does not want to be remembered or extensively broadcast. Crucially, it is also these images, projections and defences that shape the present and reconstruct the past, thus calling for an analysis of 'a dual history of illusions and actual practices'.[67] Without a self-reflexive engagement with such a dual history—or without the constant self-questioning advocated in the initial development of a deinstitutionalisation movement in the country—the legacy of such a movement encounters accusations of non-dialogue, alleged or actual practices of gate-keeping and authoritative practices that characterise an 'ivory tower'. As Segatori puts it, maintaining his theological vocabulary, such practices generate 'condemnations of heresy' to those who question a supposed 'Basaglianism'.[68]

[64] Ibid., pp. 11 and xvii.
[65] Ibid., p. 44.
[66] Ibid., p. 49.
[67] Ibid., p. xviii.
[68] Segatori, *Oltre l'Utopia Basagliana*, p. 53.

As I have articulated, the projections and defences that are intrinsic to an oscillation between algos and nostos – as well as forms of crystallisation and theoretical encapsulation implicit in a language of 'Basaglianism' – have important consequences that reverberate beyond historical discourses. Indeed, they mobilise family associations and make room for problematic definitions of Law 180 as 'an unscientific denial of mental illness and its cure, a rough and unrealistic denial of reality itself'.[69] In this respect, while Jervis is interested in developing a new vocabulary to replace the slogans and motifs that emerged in the 1960s – suggesting that the critique of institutional psychiatry needs to be re-contextualised with a degree of historical awareness – Segatori's critique of such symbols reaches drastic and troublesome tones.[70] He charges that 'if we were to take away from these more or less prestigious representatives two key words like stigma and mental asylum, which are brandished to terrorise patients and make families and society at large feel guilty, they would have nothing to cling on to'.[71]

The employment of a symbolic phraseology thus becomes a problematic aspect of 'Basaglianism', especially when coupled with the 'romances with the past' that characterise feelings and practices of nostalgia. In fact, the atmosphere of 'Basaglian' staff meetings has been described as one of marked 'sentimentalism', an 'auto-celebratory description of its own practices and successes', an 'impermeable over-consideration' of such work, within an 'atmosphere of self-referential approbation, facilitated by iconographic propaganda operations and sentimental marketing'.[72] Such sentimental iconography explicitly invests several symbols that are directly involved in the oscillations between nostos and algos, or restorative and reflective nostalgia. I here focus on one of these symbols, outlining the vicissitudes borne by Marco Cavallo.

Marco Cavallo

Commenting on the recent commotions around the nomination of the new Director of the Trieste Mental Health Department, a local journalist reports that, as a witness to such controversies, 'Marco Cavallo ... stays

[69] Ibid., p. 58.
[70] Jervis and Corbellini, *La Razionalità Negata*; Segatori, *Oltre l'Utopia Basagliana*.
[71] Segatori, *Oltre l'Utopia Basagliana*, p. 79.
[72] Ibid., pp. 26 and 94.

still and observes'.[73] Marco Cavallo ('Marco the Horse') is both a sculpture inside the park of Trieste's former psychiatric hospital, and the logo of the Trieste Mental Health Department. Decades before turning into an avatar and a metal icon, Marco Cavallo had been the living horse that used to carry the laundry around the complex of the hospital since the late 1950s. When the local administration decided to substitute him with a lorry in early 1972, he was destined for the abattoir. In June 1972, a group of patients and staff members campaigned for the horse not to be put down, and a local farmer eventually offered to give him shelter.

In January 1973, Basaglia gathered a team of artists to work with the hospital community. Artist Vittorio Basaglia and theatre practitioner Giuliano Scabia began a project that involved the collective construction of enormous puppets, establishing a workshop named Laboratorio P, where patients were invited to propose themes and ideas. By then, the horse had already left the hospital, but when a patient's drawing took the shape of a giant papier-mâché horse, the community agreed that the sculpture represented Marco Cavallo. Inside its belly, patients put notes and cards where they wrote their desires, aspirations and wishes. The sculpture was then painted light blue.

The proposal emerged, in the hospital community, to parade the horse sculpture around Trieste, accompanied by marching patients and staff members. This, however, raised a heated debate among staff and patients: there was the risk that the parade might appear to the public as a climax, an ossified and celebrative peak in the work of the humanisation of the asylum. They eventually reached a compromise, and Marco Cavallo paraded around the city, while posters and leaflets were distributed which informed the public of the current status of the hospital and the work that still lay ahead.[74]

In recounting the events, Dell'Acqua admits that memories are often entangled with fables: fictions and myths coalesce with facts in inextricable

[73] Dorigo, 'Dsm, Cambio al Vertice da Mezzina a Signorelli'.

[74] A. Sforza Tarabochia (2008) 'When the Outside Met the Inside: Marco Cavallo as a Symbol of Demarginalisation in Italian Psychiatry', Chiasmi, Harvard-Brown Postgraduate Conference, Brown University, Providence (RI), USA, 15 March; Dell'Acqua, *Non ho l'Arma che Uccide il Leone*. The assembling and parading of Marco Cavallo in 1973 is photographically documented on the Trieste Mental Health Department website, while Marco Cavallo—the stylised logo—ubiquitously appears on the website, see http://triestesalutementale.it, accessed 4 October 2013. These images were not made available for reproduction in this text.

Fig. 11.1 Metal sculpture of Marco Cavallo (former Psychiatric Hospital of Trieste, February 2012, author's collection)

ways. It seems that Marco Cavallo was too big to exit the door of the old ward in which he had been assembled. He was running up and down the corridor until he decided to knock the door down, where Basaglia played the literal part of the battering ram. Again, the external gate of the hospital was too small and had to be smashed for the horse to run freely. Marco Cavallo—the animal and the sculpture—had exited the asylum. From this blend of fact and fiction, 'Basaglia ramming down the fence of the asylum was the symbolic inaugural act' of the movement of the hospital towards the outside.[75]

The papier-mâché Marco Cavallo is now represented by a metal sculpture inside the complex of the former hospital in Trieste (Fig. 11.1). An unspecified number of reproductions of the horse have been created over the years (Fig. 11.2). Marco Cavallo represents 'the symbol of freedom from mental asylums' and, more broadly, the 'symbol of psychiatric health care in Italy'.[76] One of these sculptures conducts occasional journeys

[75] Sforza Tarabochia, 'When the Outside Met the Inside'.
[76] Ibid.

Fig. 11.2 Two reproductions of Marco Cavallo (former Psychiatric Hospital of Trieste, February 2012, author's collection)

through the country, as part of Democratic Psychiatry's ongoing campaign for the abolition of criminal psychiatric hospitals—never subjected to Law 180, since they are managed by the Interior Ministry.[77]

Symbols, Models and Permanent Crisis

Marco Cavallo has thus become—literally and symbolically—a monument to Basaglia's work in Trieste and a symbol of psychiatric deinstitutionalisation in the country. As a monument, it presents many of the issues that Paul Connerton identifies in practices of memorialisation.[78] Arising from the fear of cultural amnesia, memorials might lead to the progressive discarding of the obligation to remember. In fact, '[T]he relationship between memorials and forgetting is reciprocal: the threat of forgetting

[77] http://news-forumsalutementale.it, accessed 27 July 2013; and http://psichiatriademocratica.it, accessed 20 September 2013.

[78] P. Connerton (2009) *How Modernity Forgets* (Cambridge: Cambridge University Press).

begets memorials and the construction of memorials begets forgetting'.[79] For Connerton, institutionalised and officially sanctioned forms of hyper-remembering paradoxically generate post-mnemonic cultures, with a marked tendency towards forgetting. The issue at stake is precisely one of remembering and forgetting Basaglia, and the principles that inspired his work. Marco Cavallo—not only a monument, but also the symbol of post-deinstitutionalisation psychiatric healthcare—is precisely a token of a heritage and legacy constructed around Basaglia.

Symbols, however, have a tendency not only to crystallise the passing of time, but also to ossify what was initially conceived of as a dynamic process, in need of constant development. Marco Cavallo—the symbol, the logo and the sculpture—thus risks being severed from the communal discussions and debates that literally assembled it and took it parading around Trieste. Similarly, Basaglia's work and practices of deinstitutionalisation—founded upon constant self-questioning, dialectic confrontations and challenges to power relations—have allegedly been turned into Basaglianism—irrefutability, rigidity of thought and forms of non-dialogue. In this sense, both Basaglia and Marco Cavallo are employed as discursive currencies—albeit at different levels—that partake in nostalgic practices where the past is carried forward through symbols and reconstructed to serve present agendas. As Boym notes, in restorative nostalgia '[d]istance is compensated by intimate experience and the availability of a desired object. Displacement is cured by a return home .... Never mind if it's not your home; by the time you reach it, you will have already forgotten the difference.'[80]

Difference, indeed, has perhaps been neglected. Against the notion of permanent crisis and an 'anti-model' framework that underpinned the development of the deinstitutionalisation movement from Gorizia, and in order to defend the legacy of such movement, the Trieste Mental Health Department is now portrayed as a 'small fort', a 'bastion' or a 'model' of 'the Basaglia cure'.[81] Paradoxically, a 'model' runs against the notions of movement, disruption, subversion, dynamics and permanent crisis which

[79] Ibid., p. 29.

[80] Boym, *The Future of Nostalgia*, p. 44.

[81] F. Basaglia (1971) 'Il Problema della Gestione' in Basaglia (ed.), *L' Istituzione Negata*, pp. 370–80, 370; G. Ziani (2013) '"Niente di Personale, è l'Ass a Soffrire". Ma c'è una lettera ...', *Il Piccolo*, 9 February, at http://ilpiccolo.geolocal.it, accessed 4 October 2013; Ziani, 'Signorelli (Dsm): Ho i Titoli'; D' Alessandro, *Lo Specchio Rimosso*, p. 261; Ziani, 'Lavoro Contro il Disagio Mentale'.

characterised the initial development of the 'pioneering deinstitutionalisation experiences of Franco Basaglia and his collaborators', from which, one reads on its web page, the Trieste Mental Health Department had 'directly evolved'.[82]

If during the twentieth century Italian psychiatry 'has been constructing its past, as it constructed itself',[83] Basaglia is still currently deployed in this construction through forms of nostalgia that generate ossifying frames of reference. This issue not only poses questions around the uses of history, but becomes even more problematic as it appears to repudiate the principles that launched the movement in the first place. If the image of the Trieste Mental Health Department is one of 'coherence' 'with the philosophy that inspired the reform', I have here taken issue with such coherence, on the basis that this 'philosophy' was based on dialogue, constant self-questioning and the rejection of optimally functioning models, where 'every perfect world is an ideological deceit'.[84] As might be expected, the issue is a very sensitive one. Non-compliant inquiries and explicit critiques might produce defensive responses, refusals or belittlements that characterise practices of 'ivory towering', and they might trigger internal 'rumpuses', for there is 'nothing worse than to accuse the Basaglians of being prisoners of univocal thinking'.[85]

Conclusion

Reporting on the vicissitudes around the nomination of the new Director of the Trieste Mental Health Department, Dorigo opens his article by stating that 'Franco Basaglia was not a Basaglian'.[86] It is within this distinction that I have here juxtaposed Basaglia against Basaglia's work in deinstitutionalisation. As he analysed the developments in Gorizia, Basaglia anticipated that 'the danger now is that the new institutions will be given this new appearance, and that there will be a fixed and final arrangement, once again', and he noted that 'if our initial defiance had a political character,

[82] http://triestesalutementale.it, accessed 4 October 2013.

[83] Guarnieri, *La Storia della Psichiatria*, p. 14.

[84] Norcio, 'La Legge 180', p. 175; M. G. Giannichedda (2008) 'Introduzione' in F. Ongaro Basaglia (ed.) *L'Utopia della Realtà* (Turin: Einaudi), pp. vii–lii, xliv.

[85] Ziani, 'Signorelli (Dsm): Ho i Titoli'.

[86] Dorigo, 'Dsm, Cambio al vertice da Mezzina a Signorelli'.

the political act risks being absorbed as a technical act, and it must therefore reinvent itself as political'.[87]

Finally, in articulating one of the main problems facing the condition of the institutionalised patient, Basaglia explained that 'de-historicised, de-responsabilised, a-problematic, a-dialectic, the inmate in the psychiatric institution, however, fulfils his function of supporting the system, precisely as he is categorised and defined within one of its institutions'.[88] I wonder what the effects might be if we were to substitute in this statement 'the inmate in the psychiatric institution' with the name 'Basaglia'. If this substitution can be thought-provoking, rather than just provocative, it might cast light on the processes where revolutionary or progressive action ends and its stabilisation into a new form of knowledge begins, ossified into 'models', 'bastions' or 'towers'.

[87] Basaglia, 'Le Istituzioni della Violenza e le Istituzioni della Tolleranza', pp. 80, 85.
[88] Basaglia, 'Il Problema della Gestione', p. 375.

CHAPTER 12

A Mental Health System in Recovery: The Era of Deinstitutionalisation in California

Howard Padwa, Marcia Meldrum, Jack R. Friedman, and Joel T. Braslow

Introduction

California, like many states across the United States (US), began to reduce dramatically the population of its state hospitals in the 1950s. Between 1959 and 1973, California's state hospital census showed a decrease in patient numbers of approximately 80 per cent,[1] and policymakers anticipated that a nascent community-based system of mental health clinics would be able

[1] D. R. Kemp (1991) 'Mental Health Crisis in California', *Journal of Mental Health Administration*, 18:2, 154–64.

H. Padwa (✉) • M. Meldrum • J.T. Braslow
UCLA, Los Angeles, USA
e-mail: hpadwa@ucla.edu; meldrum@history.ucla.edu; jbraslow@mednet.ucla.edu

J.R. Friedman
University of Oklahoma, Norman, USA
e-mail: jack.r.friedman@ou.edu

D. Kritsotaki et al. (eds.), *Deinstitutionalisation and After*,
Mental Health in Historical Perspective,
DOI 10.1007/978-3-319-45360-6_12

to meet the needs of former psychiatric patients. However, as was the case elsewhere in the US,[2] the community-based mental health system did not meet the needs of the state's deinstitutionalised population; local community mental health programmes were chronically underfunded,[3] and ill-equipped to meet the array of psychiatric, social and economic needs of individuals with severe mental illness.[4] Consequently, many Californians living with severe mental illness struggled, cycling through periods of hospitalisation, residence in nursing homes, incarceration and homelessness.[5]

Moreover, as the groundbreaking work conducted by Sue Estroff in the 1970s showed, even for individuals who were able to meet their basic subsistence needs, community-based care was unable to facilitate meaningful social integration: medication side effects made it difficult to function independently; in spite of disability welfare subsidies, they continued to live in isolation from the mainstream community; and placement in supported work settings lowered their confidence and led to demoralising episodes of inactivity. Consequently, in spite of deinstitutionalisation's potential to facilitate community integration, individuals with serious mental illness living in the community nonetheless remained chronically marginalised and isolated from mainstream society.[6]

In the 40 years since these shortcomings of deinstitutionalisation have become manifest, stakeholders in the California mental health system—

[2] S. M. Rose (1979) 'Deciphering Deinstitutionalization: Complexities in Policy and Program Analysis', *Milbank Memorial Fund Quarterly. Health and Society*, 57:4, 429–60; R. I. Freedman and A. Moran (1984) 'Wanderers in a Promised Land: The Chronically Mentally Ill and Deinstitutionalization', *Medical Care*, 22:12, S1–S68; A. Scull (1985) 'Deinstitutionalization and Public Policy', *Social Science and Medicine*, 20:5, 545–52; D. Mechanic and D. A. Rochefort (1990) 'Deinstitutionalization: An Appraisal of Reform', *Annual Review of Sociology*, 16, 301–27; G. N. Grob (1995) 'The Paradox of Deinstitutionalization', *Society*, 32:5, 51–59.

[3] J. R. Elpers (1989) 'Public Mental Health Funding in California, 1959 to 1989', *Psychiatric Services*, 40:8, 799–804.

[4] California Commission for Economic Development Task Force for the Seriously Mentally Ill (1987) *An Integrated Service System for People with Serious Mental Illness: A Preliminary Proposal* (Sacramento, CA: California Commission for Economic Development); Collection of Materials on Mental Health Advocacy in California (Manuscript Collection 510); Louise M. Darling Biomedical Library History and Special Collections for the Sciences, University of California, Los Angeles.

[5] Kemp, 'Mental Health Crisis'; H. H. Goldman and J. P. Morrissey (1985) 'The Alchemy of Mental Health Policy: Homelessness and the Fourth Cycle of Reform', *American Journal of Public Health*, 75:7, 727–31.

[6] S. Estroff (1981) *Making It Crazy* (Berkeley, CA: University of California Press).

family members of individuals with mental illness, policymakers and mental health clients themselves—have tried to address these challenges in different, but interrelated ways. In this chapter, we examine how these groups have responded to deinstitutionalisation at three different points in time using a variety of different methods—analysis of historical primary source documents, oral history interviews and ethnographic observations conducted in public mental health clinics located in Los Angeles County, California over a 4-year period (2007–2011). In the chapter's first section, we utilise primary source documents and oral history interviews to tell the story of how family members trying to care for loved ones with psychiatric illness coped with the tremendous challenge of helping their relatives thrive in the community in the 1970s and early 1980s, and how they used their experiences to become advocates for system change. In the chapter's second section, we describe how advocates and policymakers attempted to improve the mental health system's shortcomings in the era of deinstitutionalisation by utilising a new service philosophy—the recovery model—to reorganise and reorient public mental health services in the early 2000s. In the chapter's third section, we explore the experience of two individuals with mental illness who received services under the recovery model, based on annual interviews they conducted with our study team and ethnographic observation. Their experiences show how many of the same real-world issues and concerns that complicated community based care for individuals with mental illness in the 1970s remained problematic for mental health clients even after major shifts in California's mental health policy and practice from the 1980s through the 2000s.

The chapter's three sections provide snapshots showing how different groups of individuals both reacted to and helped create California's response to the challenges of deinstitutionalisation. Taken together, they illustrate how deinstitutionalisation created challenges that were both personal and political for stakeholders in California's mental health system, how these issues have persisted for decades, and ways in which they continue to impact Californians living with mental illness into the present day.

Family Members Respond: Mutual Support and Advocacy

The challenge of deinstitutionalisation for families

Though large numbers of Californians with the most severe mental illnesses were left on their own in deinstitutionalisation's wake, many of them—perhaps the majority—wound up living at home with family members.[7] Though many parents and other relatives had always played an active role in caring for their loved ones with severe mental illness, the burden increased exponentially when ex-patients returned home from state hospitals. Family members were forced to fill gaps in the community-based mental health service system, intermittently serving as impromptu caregivers, system navigators, gatekeepers and advocates for their relatives, even though most of them had no clinical training and were working in full-time jobs. In addition, they needed to manage daily challenges associated with their relatives' illnesses, helping them pass through episodes when they became highly symptomatic, ensuring that they did not harm themselves or others, and protecting them from neighbours who were fearful or angry about their erratic behaviour.

When families could not manage these responsibilities, they often suffered from feelings of anxiety, anger and guilt. As one mother living with a child who had severe mental illness explained, 'nobody ever knows what to expect, the atmosphere surrounding the individual becomes tense and jumpy … one of the greatest strains is never knowing if we are doing the right thing'.[8] In particular, family members reported feeling overwhelmed since they did not know how to manage situations when their loved ones were in crisis. 'Families need training on how to cope', said one parent speaking to the California Assembly Permanent Subcommittee on Mental Health and Developmental Disabilities in 1977: 'What does one do with a suicidal person? … What are the signs of over-drugging? How hard should one push for the patient to get up and dress and undertake some activity?'[9] Exasperated and frustrated, some family members felt culpable for their loved ones' problems. 'Family members', noted advocates in 1978, 'are

[7] Comptroller General of the United States (1977) *Report to the Congress: Returning the Mentally Disabled to the Community. Government Needs To Do More* (Washington, DC: General Accounting Office), http://www.gao.gov/products/HRD-76-152, accessed 2 May 2015.

[8] P. Richardson (1991) 'Living with Mentally Ill Family Members', *AMI-LA Newsletter*, April 1991, Mental Health Advocacy Collection, University of California, Los Angeles.

[9] *Consultant Report on Improving California's Mental Health System: A Framework for Public Contributions, prepared for the California Assembly Permanent Subcommittee on Mental Health and Developmental Disabilities* (1977) Mental Health Advocacy Collection, University of California, Los Angeles.

usually guilt-ridden and feel a keen sense of failure for having "produced" a schizophrenic.'[10]

Though families were eager for help and guidance, the treatment programmes and psychiatrists who were supposed to provide community-based care were sources of frustration more often than they were helpful resources. Family members found that staff in community treatment programmes were often unprepared or unwilling to treat their relatives. 'We often hear community health workers say they "do not want to baby-sit schizophrenics"', reported a family support and advocacy group in 1975.[11] Moreover, when treatment programmes did help, families found that providers intentionally excluded them from treatment decisions and ignored their requests for help dealing with crises, finding housing or securing resources for their loved ones.

To make matters worse, many psychiatrists of the era believed that families had little to contribute to treatment and that family involvement was likely to exacerbate mental illnesses. Though psychodynamic theories of 'schizophrenogenic' motherhood that attributed the onset of schizophrenia to parenting and family dynamics[12] had been largely discredited by the 1970s, relatives found that many mental health professionals continued to treat them as part of their loved ones' problems, if not the root of them. 'Many schools of psychiatric thought hold the patient's family responsible for aggravating and even generating this illness', noted a psychiatrist and parent advocate in 1978.[13] Consequently, family members reported being purposively isolated from their loved ones' treatment. 'As parents of schizophrenics', explained one parent in 1975, 'we are told that we are too emotionally involved.'[14]

For many families, the shortcomings of the community-based mental health system had become clear by the mid-1970s: the dearth of community-based programmes capable of meeting the full spectrum of their children's psychiatric and socioeconomic needs was largely to blame

[10] R. Lamb and E. Oliphant (1978) 'Schizophrenia Through the Eyes of Families', *Hospital and Community Psychiatry*, 29, 803.

[11] P. Feerick (1975) *Parents of Adult Schizophrenics Statement*, Mental Health Advocacy Collection, University of California, Los Angeles.

[12] F. Fromm-Reichmann (1948) 'Notes on the Development of Treatment of Schizophrenics by Psychoanalytic Psychotherapy', *Psychiatry*, 11, 263–73.

[13] Lamb and Oliphant, 'Schizophrenia Through the Eyes of Families'.

[14] Feerick, *Parents of Adult Schizophrenics Statement.*

for their predicament.[15] 'Families of schizophrenics', wrote two advocates in 1978, 'have received too little help from mental health professionals, even though in many cases family are the real primary care agents for long-term patients released from mental hospitals.'[16] The issue was not only a lack of assistance, they noted, but reluctance among community programme staff to care for individuals living with the most severe mental illnesses. 'We need long-term programs staffed by persons who are not only willing to work with schizophrenics', advocates noted, 'but [who] consider it a worthwhile job'.[17]

Mutual support and advocacy

As families across America coped with the challenge of caring for loved ones with severe mental illness in the 1970s, many of them began forming mutual support and advocacy groups. In California, one of these groups formed in San Mateo County (in the San Francisco Bay Area), in 1973. Over half of the severely mentally disabled population in San Mateo lived with family members, many of whom met each other in waiting rooms when they brought their relatives to psychiatrist appointments, emergency rooms or community mental health programmes. In 1973, Eve Oliphant, a housewife and mother of an adult child living with schizophrenia, invited other parents in similar situations to meet in her home and to 'cry on each other's shoulders'. At the first meeting, she met labour organiser Tony Hoffman and his wife Fran, and together they began advocating for policies that would expand treatment options for their children. Initially, they hoped to convince policymakers to keep more state hospitals open in order to increase the availability of inpatient care. That effort soon fell short, however, so Oliphant and the Hoffmans switched focus and began advocating for improvements in community-based mental health facilities, treatment programmes and crisis care services.[18]

Oliphant and the Hoffmans officially formed a new group—Parents of Adult Schizophrenics (PAS) of San Mateo County—in 1974. Though initially brought together to provide mutual support as they coped with

[15] Ibid.

[16] Lamb and Oliphant, 'Schizophrenia Through the Eyes of Families'.

[17] Feerick, *Parents of Adult Schizophrenics Statement.*

[18] F. Hoffman (1984) 'History, 1974–1984', Mental Health Advocacy Collection, University of California, Los Angeles.

the challenges associated with schizophrenia, the group soon broadened its scope and began advocating for improvements in the services available to all people with mental illness. By early 1975, 115 parents had joined the San Mateo group, and by 1976 affiliated chapters were formed in ten neighbouring counties. Following a sit-in at the Governor's office advocating for improvements to California's mental health system, PAS began garnering significant media attention, being featured in numerous newspaper, radio and television stories. Soon thereafter, members of the group's leadership began to gain notoriety, both within the psychiatric field and in mental health policy circles. In 1976, Oliphant made a presentation at the American Psychiatric Association's annual meeting, and in 1978, she and psychiatrist Richard Lamb co-authored an article in the leading psychiatric research journal *Hospital and Community Psychiatry*.[19] In 1976, Tony Hoffman won a seat on the San Mateo County Mental Health Board, marking the first time that a family member held a formal position related to mental health policymaking in California.[20]

While PAS was gaining influence in Northern California, groups in Southern California—Los Angeles in particular—began forming their own mental health support groups for relatives of individuals living with mental illness. In 1977 and 1978, groups across the state organised to form the California Association of Families of the Mentally Disabled (CAFMD) under the leadership of Hoffman and Los Angeles parent advocates Stella March, Don Richardson and Peggy Richardson. Initially, CAFMD focused on family support, improved community awareness of mental illness and advocacy for better community treatment and housing facilities. The group also collaborated with mental health consumer groups—advocacy groups composed of individuals living with mental illness themselves—in order to push for increased funding and reform. In 1978, CAFMD, together with consumer groups, achieved a major victory with the passage of a law that required the California Citizens' Advisory Council, which advised the Governor and legislature on mental health policy, to include mental health consumers and family members.[21]

In 1979, CAFMD became integrated into a broad movement of family advocacy surrounding mental illness that had been developing across the US. In 1979, a meeting of representatives from 59 family groups led

[19] Lamb and Oliphant, 'Schizophrenia Through the Eyes of Families'.

[20] Hoffman, 'History, 1974–1984'.

[21] Ibid.

to the formation of the National Alliance for the Mentally Ill (NAMI).[22] CAFMD became a NAMI chapter, and renamed itself the California Alliance for the Mentally Ill (CAMI) in 1982. In less than a decade, a disparate collection of family support groups had organised and transformed themselves into members of a national advocacy organisation that would exert influence on both state and national mental health policies. Like its parent organisation, CAMI had two major policy goals; to reduce the stigma surrounding psychiatric conditions, and to enhance community-based services in order to help family members as they cared for their loved ones with mental illness.[23]

In its work on stigma, CAMI allied itself with psychiatrists and other mental health professionals who advocated for biological understandings of mental illness that sought to normalise it by comparing it to physical conditions such as cancer and diabetes. The concept of mental illness as being a 'real physical brain dysfunction' rather than something 'caused by faulty interpersonal relationships', explained Don Richardson in a 1988 speech, promised to both mitigate stigma and counter traditional psychodynamic theories which held that families caused mental illness.[24] CAMI simultaneously advocated for a view of mental illness that destigmatised it both for individuals who had these conditions and their family members. From 1989 through 2000, CAMI President Daniel Weisburd published the *CAMI Journal*, a periodical that featured articles describing the biology of mental illness, combating public misconceptions of mental illness, and detailing the experiences of individuals who lived with mental illness and others in their lives.[25] In 1995, March initiated a nationwide 'Stigma Busters' campaign to monitor movies, television shows and news media, and advocate for less stigmatising depictions of individuals with mental illness in movies, on television and in journalism. March also sought to humanise mental illness through her 'In Our Own Voice' video series, in

[22] National Alliance for Mental Illness (NAMI) Wisconsin (2015) 'Mission and History', http://www.namiwisconsin.org/mission-history/, accessed 6 July 2015.

[23] Hoffman, 'History, 1974–1984'.

[24] D. Richardson (1988) 'AMI-LA Keynote Address, 23 June 1988', Mental Health Advocacy Collection, University of California, Los Angeles.

[25] D. Weisburd (2010) Oral History Interview conducted by H. Padwa and K. Miller, 2010. *Hope Story: History of Public Mental Health Care in California*, http://histpubmh.semel.ucla.edu/sites/default/files/oral_histories/3974b795_.pdf, accessed 6 July 2015.

which celebrities living with mental disorders and ordinary patients shared their personal experiences.[26]

Controversies and successes

In its work to aid family members by enhancing community-based services, CAMI sometimes came into conflict with consumer groups and even its own members. In particular, many CAMI members advocated for policies to facilitate involuntary treatment of individuals with mental health disorders in crisis situations. The group's position on involuntary treatment had its roots in family members' experiences when they had difficulty controlling their relatives' behaviours, and could not help loved ones who refused psychiatric treatment. CAMI members pushed for state laws (such as 'Kendra's Law' in New York) that would shift the burden of crisis care to professionals by loosening involuntary commitment laws and allowing for forced treatment. In this area, CAMI encountered significant opposition from consumer groups who considered involuntary treatment a violation of civil rights. The issue also caused major schisms within CAMI itself; in 2000, Weisburd, who was an ardent opponent of forced treatment, suddenly resigned his post as CAMI president in the wake of internal disputes related to involuntary treatment.[27]

In spite of these controversies, CAMI had a tremendous influence on the development of California's mental health policy beginning in the 1980s. In response to complaints from family members, California Lieutenant Governor Leo McCarthy convened a task force—headed by Weisburd—to identify ways that the state's mental health system could be improved.[28] The task force concluded that the community-based system of care that had emerged during deinstitutionalisation was woefully inadequate, and that new service delivery models designed to provide com-

[26] S. March (2009) Oral History Interview conducted by M. Meldrum and A. Nelligan, 2009. *Hope Story: History of Public Mental Health Care in California*, http://histpubmh.semel.ucla.edu/sites/default/files/oral_histories/c890d394_Stella_March_.pdf, accessed 6 July 2015.

[27] Weisburd, Oral History Interview.

[28] Ibid; California Commission for Economic Development Task Force for the Seriously Mentally Ill (1987) *An Integrated Service System for People With Serious Mental Illness: A Preliminary Proposal* (Sacramento, CA: California Commission for Economic Development Task Force for the Seriously Mentally Ill), Mental Health Advocacy Collection, University of California, Los Angeles.

prehensive 24-hour care could help address the needs of individuals with severe mental illness living in the community.[29] In response to the task force's findings, the California legislature approved Assembly Bill (AB) 3777 in 1988, authorising the creation of Integrated Service Agencies (ISAs), pilot programmes that followed the task force's recommendations.[30] Based on pilot results, several counties established their own ISA programmes, and the California legislature approved funding to expand the ISA model in 1999 and 2000.[31]

Thus by the turn of the new millennium, family members of individuals living with severe mental illness had become powerful players in the arena of California mental health policy, and they were instrumental in facilitating the creation and expansion of services designed to meet the needs of individuals with severe mental illness living in the community. Having faced tremendous challenges caring for their loved ones in the aftermath of deinstitutionalisation, family members turned to mutual support and then advocacy, and helped spearhead a wave of reform that would eventually bring about major changes to California's mental health system in the early twenty-first century.

Policymakers Respond: Empowerment and Recovery

The emergence of the recovery model

While family members had direct experience with the shortcomings of deinstitutionalisation almost immediately, policymakers were slower to respond to the epidemics of homelessness, incarceration and social marginalisation among the deinstitutionalised population. The creation of the National Institute of Mental Health's Community Support Program (CSP) in the late 1970s began to address many of the challenges of deinstitutionalisation by creating pilot programmes that provided an array of psychosocial services and support (e.g. case management, housing assistance, living skills training) in addition to psychiatric care,[32] and the ISAs

[29] Task Force for the Seriously Mentally Ill, *An Integrated Service System.*

[30] State of California (1988) Assembly Bill 3777.

[31] State of California (1999) Assembly Bill 34; State of California (2000) Assembly Bill 2034.

[32] J. C. Turner and W. J. TenHoor (1978) 'The NIMH Community Support Program: Pilot Approach to a Needed Social Reform', *Schizophrenia Bulletin* 4:3, 319–49.

in California offered a similar array of comprehensive services to help individuals with severe mental illness live in the community.[33]

Yet in the late 1980s and early 1990s, a growing chorus of clients, providers, researchers and policymakers began arguing that these services did not do enough to help individuals with mental illness actually *thrive* in the community. According to these critics, the main issue was not the scarcity of mental health services, but rather their orientation. Proponents of what eventually came to known as the 'recovery model' for mental health care argued that psychiatric care usually utilised 'disempowering treatment practices' that perpetuated a cycle of hopelessness and isolation for individuals with mental illness.[34] Recovery advocate Patricia Deegan—a psychologist and mental health client herself—asserted that both pharmacological and behavioural services offered in treatment settings robbed mental health clients of their capacity and initiative to function independently: psychiatrists, she claimed, doled out soul-sucking medications that 'made us slur and shake, that robbed our youthful bodies of energy, and made us walk like stiff zombies', thus 'radically diminish(ing) our personhood and sense of self',[35] while therapists encouraged clients to identify with their illnesses, and avoid activities that went above and beyond the narrowly constructed social confines of an appropriate 'mental patient'. Consequently, recovery advocates maintained, mainstream mental health services had effects that were both 'iatrogenic'[36] and 'spirit breaking',[37] and were apt to cause as much suffering as psychiatric disorders themselves, if not more.

Advocates of recovery-oriented reform believed that by realigning mental health services to focus more on cultivating independence and individualism, they could help individuals living with mental illness fully

[33] Lewin-VHI, Inc., J. Meisel and D. Chandler (1995) *AB 3777 System Reforms: The Integrated Service Agency Model. A Summary Report to the California Department of Mental Health* (Sausalito, CA: Lewin-VHI), Mental Health Advocacy Collection, University of California, Los Angeles; State of California, Assembly Bill 34; State of California, Assembly Bill 2034.

[34] W. Anthony (1993) 'Recovery from Mental Illness: The Guiding Vision of the Mental Health Service System in the 1990s', *Psychiatric Rehabilitation Journal*, 16:4, 11–23.

[35] P. Deegan (1987) *Recovery, Rehabilitation, and the Conspiracy of Hope.* Available online at https://www.patdeegan.com/sites/default/files/files/conspiracy_of_hope.pdf, accessed 1 September 2015.

[36] Anthony, 'Recovery from mental illness'.

[37] Deegan, *Recovery, Rehabilitation, and the Conspiracy of Hope.*

recover and thrive as full members of the community.[38] Throughout their writings and speeches, advocates framed recovery as a corrective to what they viewed to be the disempowering nature of mental health care in the 1980s and 1990s: whereas the prevailing mental health system inculcated hopelessness and discouraged clients from pursuing a better future, a recovery-oriented mental health system would foster hope and kindle a desire to pursue their dreams;[39] while the existing mental health system simply encouraged clients to remain medication-adherent and do as they were told,[40] a recovery-oriented mental health system would give clients the right to choose the course of their treatment and pursue whatever life goals they chose;[41] traditional mental health care focused on identifying and managing functional deficits,[42] whereas a recovery-oriented system of care would be based on identifying and working from clients' strengths;[43] and while the existing mental health system discouraged clients from taking risks in order to improve their lives, a recovery-oriented mental health system would grant them the 'dignity of risk' and 'right to failure' that was accorded to everyone else in US society.[44]

[38] Anthony, 'Recovery from mental illness'; L. Davidson, M. O'Connell, J. Tondora, T. Styron and K. Kangas (2006) 'The Top Ten Concerns about Recovery Encountered in Mental Health System Transformation', *Psychiatric Services*, 57:5, 640–45.

[39] Deegan, *Recovery, Rehabilitation, and the Conspiracy of Hope*; L. Davidson and D. Roe (2007) 'Recovery from versus Recovery In Serious Mental Illness: One Strategy for Lessening Confusion Plaguing Recovery', *Journal of Mental Health*, 16:4, 459–70; Substance Abuse and Mental Health Services Administration (2006) *From Exclusion to Belonging: Transforming Mental Health Care in America* (Rockville, MD: US Department of Health and Human Services).

[40] Deegan, *Recovery, Rehabilitation, and the Conspiracy of Hope*; E. Leete (1989) 'How I Perceive and Manage My Mental Illness', *Schizophrenia Bulletin*, 15:2, 197–200.

[41] Deegan, *Recovery, Rehabilitation, and the Conspiracy of Hope*; R. Ralph (2000) 'Recovery', *Psychiatric Rehabilitation Skills*, 4:3, 480–517; President's New Freedom Commission (2003) *Achieving the Promise: Transforming Mental Health Care in America: Final Report* (Bethesda, MD: US Department of Health and Human Services).

[42] S. J. Onken, C. Craig, P. Ridgway, R. Ralph and J. Cook (2007) 'An Analysis of the Definitions and Elements of Recovery: A Review of the Literature', *Psychiatric Rehabilitation Journal*, 31:1, 9–22.

[43] L. Davidson et al., 'Top Ten Concerns'.

[44] Deegan, *Recovery, Rehabilitation, and the Conspiracy of Hope*; P. Deegan (1988) 'Recovery: The Lived Experience of Rehabilitation', *Psychosocial Rehabilitation Journal*, 11:4, 11–19; Ralph, 'Recovery'.

Recovery as policy in California: the Mental Health Services Act

With its focus on empowerment, independence and individualism, recovery resonated with the ideals of neo-liberal citizenship[45]—particularly those concerning self-responsibility and self-sufficiency—that had been ascendant in US health[46] and social welfare[47] policy since the Reagan administration. A natural fit in the era of neo-liberalism, recovery emerged as the main philosophical driver of mental health policy reform in the US by the end of the 1990s. In 1999, the US Surgeon General recommended that mental health services should be designed with a 'recovery perspective'[48] and in 2003, President George W. Bush's New Freedom Commission on Mental Health announced that 'recovery' should be 'the goal' of mental health system transformation across America.[49]

[45] J. Braslow (2013) 'The Manufacture of Recovery', *Annual Review of Clinical Psychology*, 9, 781–809; A. B. Adeponle, R. Whitley and L. J. Kirmayer (2012) 'Cultural Contexts and Constructions of Recovery' in A. Rudnick (ed.) *Recovery of People with Mental Illness: Philosophical and Related Perspectives* (Oxford: Oxford University Press), 109–32; N. Myers (2010) 'Culture, Stress and Recovery from Schizophrenia: Lessons from the Field for Global Mental Health', *Culture, Medicine, and Psychiatry*, 34:3, 500–28.

[46] S. McGregor (2001) 'Neoliberalism and Health Care', *International Journal of Consumer Studies*, 25:2, 82–89; S. Nkansah-Amankra, S. K. Agbanu and R. J. Miller (2013) 'Disparities in Health, Poverty, Incarceration, and Social Justice among Racial Groups in the United States: A Critical Review of Evidence of Close Links with Neoliberalism', *International Journal of Health Services*, 43:2, 217–41; J. Banda (2014) 'Rapid Home HIV Risk Testing and the Moral Imperatives of Biological Citizenship', *Body & Society*, 21:4, 24–47; E. D. Carter (2015) 'Making the Blue Zones: Neoliberalism and Nudges in Public Health Promotion', *Social Science & Medicine*, 133, 374–82.

[47] N. Fraser and L. Gordon (1994) 'A Genealogy of Dependency: Tracing a Keyword of the US Welfare State', *Signs*, 19:2, 309–36; R. Jayakody, S. Danziger and H. Pollack (2000) 'Welfare Reform, Substance Use, and Mental Health', *Journal of Health Politics, Policy and Law*, 25:4, 623–52; S. Morgen (2001) 'The Agency of Welfare Workers: Negotiating Devolution, Privatization, and the Meaning of Self-Sufficiency', *American Anthropologist*, 103:3, 747–61; Y. Hasenfeld and E. Garrow (2012) 'Nonprofit Human-Service Organizations, Social Rights, and Advocacy in a Neoliberal Welfare State', *Social Service Review*, 86:2, 295–322; M. Abramovitz (2006) 'Welfare Reform in the United States: Gender, Race and Class Matter', *Critical Social Policy*, 26:2, 336–64; S. F. Schram, R. C. Fording and J. Soss (2008) 'Neo-Liberal Poverty Governance: Race, Place and the Punitive Turn in US Welfare Policy', *Cambridge Journal of Regions, Economy and Society*, 1, 17–36.

[48] Surgeon General of the United States (1999) *Mental Health: A Report of the Surgeon General* (Rockville, MD: US Department of Health and Human Services), xvi.

[49] President's New Freedom Commission, *Achieving the Promise*.

In California, recovery became the focus of mental health policy with the passage of Proposition 63, the Mental Health Services Act (MHSA), in November 2004. In addition to raising significant funds for public mental health services across the state (it raised public mental health budgets by approximately one third),[50] one of the MHSA's main aims was to reorient mental health services to become more focused on 'recovery' as defined by advocates of the recovery model. The legislation amended the California Welfare and Institutions Codes to stipulate that 'planning for (mental health) services shall be consistent with the philosophy, principles, and practices of the Recovery Vision for mental health consumers' by promoting 'concepts key to recovery for individuals who have mental illness: hope, personal empowerment, respect, social connections, self-responsibility, and self-determination'.[51]

The centrepieces of the MHSA were Full Service Partnerships (FSPs), programmes like CSP pilots and the ISAs, which were designed to deliver comprehensive, field-based psychiatric, psychosocial, housing and rehabilitation services.[52] FSPs were explicitly designed to deliver intensive and holistic care, and the State mandated that they do 'whatever it takes' to assist their clients.[53] FSP providers were charged with either linking clients to services and benefits they needed, or providing them directly, and they were required to be available to clients 24 hours a day, seven days a week. Furthermore, FSP clients had access to flexible funding to assist with housing, food and other socioeconomic needs.[54] However, the services provided by FSPs were not designed to be permanent. Rather, policymakers envisioned them as temporary supports to help clients stabilise their lives and become more capable of living independently. By naming the FSP programmes 'partnerships', California Department of Mental Health

[50] R. Scheffler and N. Adams (2005) 'Millionaires and Mental Health: Proposition 63 in California', *Health Affairs: Millwood* W5212–W5224; Braslow, 'The Manufacture of Recovery'.

[51] California Mental Health Services Act (2005).

[52] R. Selix (2010) Oral History Interview conducted by H. Padwa, 2010. *Hope Story: History of Public Mental Health Care in California*, http://histpubmh.semel.ucla.edu/sites/default/files/oral_histories/2342b8c3_Selix_Rev_for_Web.pdf, accessed 6 July 2015.

[53] California Department of Mental Health (2005) *Mental Health Services Act Community Services and Supports: Three-Year Program and Expenditure Plan Requirements, Fiscal Years 2005–06, 2006–07, 2007–08*, Mental Health Advocacy Collection, University of California, Los Angeles.

[54] California Department of Mental Health (2005) *Mental Health Services Act Community Services and Supports.*

Director Stephen Mayberg explained, the State sought to shift the focus of services 'from "we're going to take care of you" to "we're going to work together to get you where you want to go"'.[55] As clients' mental health improved, policymakers envisioned that they would reach a point where they would no longer need such intensive services. FSP clients would, as Mayberg elaborated, learn to 'manage their illness', so that eventually 'mental health problems [would] become [their] responsibility, not somebody else's responsibility'.[56] MHSA planners' expectation was that clients would be able to 'move on' to receive less intensive care than that offered in FSPs, or no longer need mental health services at all.[57]

MHSA implementation in Los Angeles County

In Los Angeles County, the passage of the MHSA coincided with a move to reorient local mental health programmes towards the recovery model. In August 2004, the Los Angeles County Department of Mental Health (LACDMH) issued a policy mandating that the 'recovery model' be integrated into all services. According to the Department, this would require a fundamental 'shift' in service philosophy, making treatment more focused on helping clients achieve 'self-determination' than ever before.[58] The MHSA, which also focused on recovery, naturally fit with LACDMH's vision; when the legislation passed, the Department enthusiastically proclaimed that it would be a 'perfect match' for the recovery-oriented transformation that was getting underway in Los Angeles County.[59]

LACDMH policymakers estimated that there were over 112,000 Los Angeles County residents who would have benefited from enhanced services through the MHSA, but they were only able to secure funding for

[55] S. Mayberg (2013) Oral History Interview conducted by H. Padwa, 2013. *Hope Story: History of Public Mental Health Care in California*, http://histpubmh.semel.ucla.edu/sites/default/files/oral_histories/Mayberg%20oral%20history.pdf, accessed 6 July 2015.

[56] Mayberg, Oral History Interview.

[57] Ibid.; California Coalition of Mental Health Agencies (2003) *Possible Ballot Initiative: Summary of Scope and Cost of Services, 2003*, Mental Health Advocacy Collection, University of California, Los Angeles.

[58] Los Angeles County Department of Mental Health (2004) *Policy 202.28: Recovery Model*, Mental Health Advocacy Collection, University of California, Los Angeles.

[59] Los Angeles County Department of Mental Health (2004) 'Recovery Model Policy: A Perfect Match for Proposition 63', *Direct eNews*, 4 November 2004, Mental Health Advocacy Collection, University of California, Los Angeles.

4,333 FSP slots in the first three years of MHSA implementation.[60] Thus to maximise the impact of MHSA funding, LACDMH sought to utilise its FSP resources as effectively and efficiently as possible. In practice, this meant designing FSP services to facilitate client flow from the intensive (and expensive) FSP programmes to lower levels of care, and eventually out of the mental health system. The policymakers and stakeholders who designed LACDMH's plan for the MHSA envisioned that the funding would be used not just to support more comprehensive services, but also to 'focus on supporting individuals through the system toward wellness and recovery'.[61] Thus FSPs would be designed, as a 2005 policy planning document explained, to 'move folks to less intensive services and supports' over time rather than providing ongoing and unconditional care.[62] Facilitating what policymakers termed 'flow' through the mental health system was central to this vision of recovery-oriented transformation. 'The reason for transforming', summarised one document used to educate LACDMH providers about the MHSA, 'is not necessarily to achieve greater effectiveness, but to achieve greater flow.'[63]

Recovery provided LACDMH policymakers with the language and the logic needed to justify and communicate the need for client flow from FSPs to lower levels of care and eventually out of the mental health system. In particular, the argument that mental health care was disempowering and the belief that independence and self-sufficiency could bring fulfilment to mental health clients were often used by policymakers as they implemented the system-wide transformation of services under the MHSA. As one LACDMH consultant explained, psychiatric professionals in the traditional mental health system communicated 'self-reinforcing'

[60] Los Angeles County Department of Mental Health (2005) *The Los Angeles County Mental Health System's Community Services and Supports Plan: A Proposal to the California Department of Mental Health in Accordance with the Mental Health Services Act*, Mental Health Advocacy Collection, University of California, Los Angeles.

[61] Los Angeles County Department of Mental Health (2005) *Draft Worksheet for Recommendations under the Community Services and Supports (CSS) Plan for L.A. County, Adults Aged 26–59*, Mental Health Advocacy Collection, University of California, Los Angeles.

[62] Ibid.

[63] M. Ragins (2007) 'The Power of Flow' in M. Ragins, *Building Mental Health Services Act Programs* (Los Angeles, CA: Mental Health Association of Los Angeles); Collection of Materials on Mental Health Advocacy in California (Collection Number 510), Louise M. Darling Biomedical Library History and Special Collections for the Sciences, University of California, Los Angeles.

messages that psychiatric conditions are inherently disabling, and that individuals with mental health disorders are incapable of making their own decisions. Thus mental health clients were 'carefully taught that passivity and obedience is the way (to) stay out of trouble' until their spirits were 'broken' and they became completely submissive. 'Through the kindness of taking care of them', mental health providers would 'take all their [clients'] belief in themselves away'.[64] Consequently, mental health clients were, as one LACDMH publication explained, traditionally placed in 'the role of victim', completely reliant on mental health professionals to be their 'rescuer'. The mental health system, therefore, had taken on a role of 'controlling people' rather than curing them.[65] LACDMH policymakers believed that making services more focused on inculcating independence and self-sufficiency could correct these shortcomings of community-based care since deinstitutionalisation; as one consultant explained, providers needed to assert that 'we're not going to help them avoid their responsibilities anymore ... We're going to help them hold together so they *can* meet their responsibilities' to themselves and society as full members of the community.[66]

The transformation of LACDMH services with MHSA implementation reflected policymakers' concerns about making care more oriented toward encouraging independence and self-sufficiency; policymakers worried that FSP services could inadvertently 'foster a lot of dependence' on the mental health system.[67] The solution lay in creating 'Wellness Centers'—programmes that could serve as 'steps down' from FSPs and into the community. In particular, Wellness Centers were envisioned as places that would serve clients 'who are ready to take increasing responsibility for their own wellness and recovery',[68] and provide them with the training and skill development they would need to leave the mental health

[64] M. Ragins (2010) Oral History Interview conducted by M. Meldrum and H. Padwa, 2010. *Hope Story: History of Public Mental Health Care in California*, http://histpubmh.semel.ucla.edu/sites/default/files/oral_histories/78f9971c_Ragins_Oral_History.pdf, accessed 6 July 2015.

[65] Los Angeles County Department of Mental Health, 'Recovery Model Policy: A Perfect Match'.

[66] Ragins, Oral History Interview.

[67] C. Warner (2011) Oral History Interview conducted by M. Meldrum, 2011. *Hope Story: History of Public Mental Health Care in California*, http://histpubmh.semel.ucla.edu/sites/default/files/oral_histories/aaaef4e6_.pdf, accessed 6 July 2015.

[68] Los Angeles County Department of Mental Health (2006) *Approval to Further Implement the Mental Health Services Act Community Services and Supports Plan for Directly*

system and seek gainful employment.[69] Wellness Centers, explained one policy document, would be designed to 'reduce reliance on the mental health system' and prepare clients to flow out of the mental health system all together.[70]

As clinic directors began incorporating FSP and Wellness services in their clinics, they explicitly mentioned facilitating client flow from FSPs to Wellness, and eventually out of the public mental health system, as goals of their clinic transformations. Redesigned programmes, explained one clinic director, would encourage clients to 'move through the program to the greatest extent possible'. For recovery-oriented transformation to succeed, summarised another clinic director, it was 'essential' for clients to 'exit the system'.[71] Thus the 'ultimate goal' of transformed services was not to minimise symptoms or assist clients, concluded another clinic director, but 'to reduce reliance on the mental health system'.[72] Increasing client self-sufficiency, therefore, became the main focus of care. Ensuring that clients begin 'taking responsibility for themselves', explained one clinic director, was a key 'principle' undergirding the entire transformative process.[73] At many clinics, staff were instructed to operationalise these principles by measuring client progress with an assessment tool—the Milestones of Recovery Scale (MORS)—that appraises client engagement in recovery and their progress toward becoming 'self-responsible' and 'completely self-supporting'.[74]

Operated Wellness Centers and Approval of Request for Appropriation Adjustment for Fiscal Year 2006–2007, Mental Health Advocacy Collection, University of California, Los Angeles.

[69] Los Angeles County Department of Mental Health (2005) *Arcadia Mental Health Center, Family Wellness Center Proposal Draft*, Mental Health Advocacy Collection, University of California, Los Angeles; Los Angeles County Department of Mental Health (2005) *Edelman Westside Mental Health Center, ACT Team Proposal*, Mental Health Advocacy Collection, University of California, Los Angeles.

[70] Los Angeles County Department of Mental Health, *Approval to Further Implement.*

[71] Los Angeles County Department of Mental Health (2005) *South Bay Mental Health Center, Clinic Transformation Draft*, Mental Health Advocacy Collection, University of California, Los Angeles.

[72] Los Angeles County Department of Mental Health, *Arcadia Mental Health Center.*

[73] Los Angeles County Department of Mental Health (2005) *Hollywood Mental Health Center, Adult Systems Transformation: A Recovery Model*, Mental Health Advocacy Collection, University of California, Los Angeles.

[74] D. Pilon and M. Ragins (2007) 'Milestones of Recovery Scale' in Ragins, *Building Mental Health Services Act Programs*; Los Angeles County Department of Mental Health (2005) *Compton Mental Health Center, Phase I Adult Systems Transformation*, Mental Health Advocacy Collection, University of California, Los Angeles; Los Angeles County

Whereas family members in the 1970s and 1980s saw the potential cures for deinstitutionalisation's ills in expanding and strengthening the mental health system, advocates and policymakers in the 1990s and 2000s believed that the key to improving lives for individuals living with mental illness in the community was to redesign services in order to make them more geared towards facilitating self-sufficiency, independence and decreased reliance on mental health services. Yet evidence suggests that the goal of facilitating decreased reliance on the mental health system has been difficult to achieve for most clients. Data from fiscal years 2010–2011 and 2011–2012 show that nearly three quarters of Los Angeles County clients who received FSP services remained in FSP programmes instead of flowing to lower levels of care, while under nine per cent of them successfully made transitions to less intensive treatment programmes or left the mental health system.[75] As policymakers concluded in a 2012 report, 'concern for continued (client) success at lower levels of care and resistance to ending/transitioning a positive therapeutic relationship' made client transitions difficult to achieve, while concerns about insufficient financial supports for clients outside of FSP programmes also inhibited client flow.[76]

The Challenge of Flow: The Cases of Linda and Mila

The experiences of two FSP clients who flowed 'through' the LACDMH system of care further elucidate why flow has been difficult to achieve. Both Linda and Mila (pseudonyms) entered a LACDMH FSP programme within a few months of each other, and had histories of homelessness and psychiatric hospitalisation; both received similar care from the FSP team (they had the same case manager and the same psychiatrist), and both utilised a similar array of FSP services (temporary housing, counselling,

Department of Mental Health (2007) *Recovery-Based System Transformation: Organizational Change Domains and Goals, Big 13 Domains of Change Status Report*, Mental Health Advocacy Collection, University of California, Los Angeles.

[75] Los Angeles County Department of Mental Health (2011) *Mental Health Services Act Annual Update, Fiscal Year 2011–2012*; Los Angeles County Department of Mental Health (2012) *Mental Health Services Act Annual Update, Fiscal Year 2012–2013*; Los Angeles County Department of Mental Health (2013) *Mental Health Services Act Annual Update, Fiscal Year 2013–2014*.

[76] Los Angeles County Department of Mental Health (2011) *Mental Health Services Act Annual Update*.

case management). Both of them were able to achieve increased levels of independence over the course of receiving FSP services for four years, though neither of them was able to fully disengage from the public mental health system, as envisaged by policymakers.

Linda

Linda entered the FSP programme suffering symptoms of post-traumatic stress disorder (due to a long history of sexual abuse), debilitating depression, bipolar disorder and an extensive history of methamphetamine use. When she entered the FSP programme, Linda was dishevelled and dirty, having lived on the streets for many years. After four years of treatment in the FSP programme she had made significant progress; she was living independently, had maintained her sobriety for several years, and shifted from being someone who was a 'heavy burden' on the public mental health system into an individual who only occasionally received medication management services and participated in Wellness groups. Yet Linda remained enrolled in the FSP programme in spite of her progress, and in interviews she reported that she did not know what she wanted to do with herself next. She recognised the successes and improvements she had made over the previous years, but she also acknowledged that she had not yet become an independent or 'productive' person:

> I am pretty much stable. I don't have to worry about food, I don't have to worry about clothing or shelter or care or transportation. I have all that now ... But now it's going to be so that I don't destroy the future. I don't want to go backwards. I want to keep going forward .... [Q: What's 'going forward' for you?] ... getting rid of my past, the baggage. I guess now it would mean working my way into some schooling or a job, you know. I don't want to say necessarily bettering myself because I feel like I've done that a lot, but it's still bettering myself as far as education and society wise. I want to be a more active person in society, and ... have my life mean something.[77]

Linda reported that she kept busy by spending time with a close circle of friends and going to church, but she recognised that she had not found a way to 'better herself' and become a 'more active person in society', to give her life 'meaning'. As Linda explained in her third interview with the

[77] Interview Conducted for United States National Institute of Mental Health Grant R01MH080671.

research team, her mental health symptoms had become less of a burden for her, but she did not experience any newfound sense of independence, self-actualisation or security as a member of the community without support from the mental health system. In particular, Linda reported that fear about becoming fully 'recovered' was holding her back from making steps to completely leave her involvement in the public mental health system behind. In particular, she believed that pursuing an education, finding a job and completely cutting herself off from the public mental health system would expose her to serious economic risk.

If she secured full employment, Linda explained, it would likely have been a low wage, working-class job that would have offered her none of the psychosocial and socioeconomic support that she was entitled to as a person who received disability payments by virtue of her mental illness. Conversely, by remaining engaged in the public mental health system (and therefore eligible for disability support), Linda was assured that she would have the money needed to pay for rent and food, while having uninterrupted access to medical and psychiatric care. Thus, even though she had followed the path of recovery as envisioned by policymakers to a degree, the socioeconomic challenges Linda would have faced if she 'flowed out' of the public mental health system made her reluctant to pursue full 'independence'.

Mila

Whereas Linda was street tough and savvy when she entered the FSP programme, Mila was shy, quiet, and in the words of her case manager, 'almost childlike' when she began receiving FSP services. Mila's diagnosis was bipolar disorder with psychotic features, and she came to the FSP after several people—medical staff who had worked with her at the local hospital, social welfare workers and her mother—had urged her to seek treatment in the FSP programme.

Mila's history of care and support was more complicated than Linda's. She had regularly resisted taking medications she had been prescribed, leading to significant behavioural problems and occasional psychiatric hospitalisations. With support from the FSP programme, Mila's condition stabilised and she thoroughly enjoyed living in the group housing that the FSP had secured for her. Mila found the bustling atmosphere of the group housing facility to be invigorating and staff referred to her as a 'social butterfly' who was friendly and comfortable living in an environment with other individuals who faced psychiatric challenges similar to her own. Because she was comfortable in the group home, Mila resisted attempts by the FSP team to secure independent living for her. After over three years in treatment, FSP staff finally convinced Mila to move in to her own apartment. When Mila agreed to the move, FSP staff celebrated this accomplishment as a great milestone on her 'road to recovery'.

Mila, however, remained ambivalent about her new living situation, and she did not feel that independent living was a significant marker of her recovery. Rather, Mila reported high levels of stress after the move, stating in interviews that she felt less secure, less safe and more alone than before when living independently. One of her major complaints was that she felt isolated from other people and that she 'didn't have anything to do'. Though she acknowledged that there were benefits to independent living (e.g. the freedom to have her boyfriend spend the night with her), Mila also reported the fear that she would lose her housing if she misbehaved, concern that she would mismanage her money and apprehension that she would lose her disability benefits if she discontinued mental health services.

Moreover, in spite of over three years of treatment built around narratives that emphasised the importance of her 'recovery' outside of the mental health system, Mila remained unclear about how to pursue it further on her own. In an interview, she said:

> Recovery is like, for me, it would be maybe even holding down a job, you know? I think, yeah … I would like to be an actress. Well, actually, I don't want to be an actress. I want to be a celebrity, which I guess is different because sometimes you're doin' movies, or sometimes you're not doin' anything. You're just getting paid to just … to appear. And so, um, I think that's what I want. I want to have a lot of people around me … it's just that it's hard for me to reach out to people, so that's why I want to be a celebrity.[78]

Though Mila was able to speak about her next steps in recovery, her ideas about what that would entail were both unrealistic and vague (being an actress, being a celebrity), and she had no idea how to pursue her goals on her own. Though Mila achieved more stable housing in the community and greater independence from the mental health system in some

[78] Interview Conducted for United States National Institute of Mental Health Grant R01MH080671.

respects, she was not prepared to become the independent, autonomous, self-sufficient citizen envisioned by policymakers when they designed the mental health system to become more 'recovery-oriented'.

Recovery and its discontents

The cases of Linda and Mila illustrate the challenges of transitioning clients from high levels of care to low levels of care, and eventually making them independent of the mental health system. FSP services helped stabilise both of them, bringing them from a state of chronic risk (homelessness, frequent hospitalisation) to more safe and secure life circumstances, and from a life where they spent most of their time receiving mental health care, to one where they spent most of their time living in the community. In these respects, both Linda and Mila appear to have made significant strides towards 'recovery'. Yet after four years of intensive support from their FSP programmes, neither Linda nor Mila were prepared to flow out of the mental health system as anticipated by the engineers of the recovery-oriented system transformation.

In both cases, the barriers to full independence from the mental health system were not clinical; they were social. Life in the community—which in a recovery-oriented schema of flow was supposed to provide opportunities for individuals like Linda and Mila to become socially integrated and self-supporting—did not offer any clear paths for them to follow. In particular, the risks of trying to survive without the disability income support that came with care from the public mental health system were barriers that made Linda and Mila reluctant to pursue complete 'independence' from mental health care. Like other health and social service recipients who were ideally supposed to be 'empowered' by neo-liberal reforms, Linda and Mila had difficulty transcending the social, economic and cultural factors that make it difficult for many vulnerable populations to thrive without significant support.[79] The absence of what Benjamin Henwood and Robert Whitley term a 'recovery-oriented society'[80]—one that offers ample protection, support and opportunities for individuals

[79] Nkansah-Amankra, Agbanu and Miller, 'Disparities in Health'; Jayakody, Danziger and Pollack, 'Welfare Reform, Substance Use, and Mental Health'; Abramovitz, 'Welfare Reform in the United States'; Schram, Fording and Soss, 'Neo-Liberal Poverty Governance'.

[80] B. Henwood and R. Whitley (2013) 'Creating a Recovery-Oriented Society: Research and Action', *Australian and New Zealand Journal of Psychiatry*, 47:7, 609–10.

with severe mental illness outside of the specialty care system—precluded them from pursuing goals of self-sufficiency and independence as envisioned under recovery-oriented transformation.

Both Linda and Mila found the therapeutic relationships they had with their providers to be sources of comfort and warmth. Consequently, disengagement from relationships with mental health providers was not an expression of independence for Linda and Mila, but rather an unnatural isolation from relationships they found to be both curative and supportive. Instead of finding meaning and fulfilment in the community outside of the mental health system, both Linda and Mila missed the camaraderie and security of the clinic itself, even though they had theoretically achieved many of the markers of 'independence' and 'recovery'.

Conclusion

California's transformation of its mental health system through the recovery model and the MHSA was the culmination of a decades-long process of reform and transformation designed to help Californians living with mental illness thrive in the community. It expanded comprehensive psychosocial services and support designed to help individuals with mental illness become truly integrated into society, and it also sought to reorient services in order to help clients become progressively less reliant on support from the public mental health system. Yet the ideals of the recovery vision were often irreconcilable with real-world realities: increased funding from the MHSA, though significant, was not nearly enough to meet the high levels of unmet need for mental health services; belief that individuals with mental illness should strive for increased independence from public mental health care were contradicted by the fact that many clients needed the mental health system to provide them with material and clinical support for extended periods of time; and hopes of mental health clients becoming fully integrated members of society proved difficult to realise for individuals who lacked the resources needed to live well without the support provided by the mental health system. Consequently, our research indicates—in line with other work examining the experience of individuals receiving recovery-oriented care—that ideals of recovery may be extremely difficult to obtain for many clients, and inadvertently cause more stress than fulfilment.[81]

[81] J. Jenkins and E. Carpenter-Song (2005) 'The New Paradigm of Recovery from Schizophrenia: Cultural Conundrums of Improvement Without Cure', *Culture, Medicine, and Psychiatry*, 29:4, 379–413; Myers, 'Culture, Stress, and Recovery'.

Thus in spite of progress, many of the same challenges that individuals with severe mental illness faced in the early 1970s remain major issues for mental health clients, family members and policymakers today: inadequate mental health service availability, difficulty surviving in the community without intensive income and psychosocial support, and persistent social marginalisation. All remain as significant for individuals like Linda and Mila today as they were for the clients described in Estroff's research that was conducted in the 1970s. At various points over the past four decades, different groups have responded to these challenges in different ways. For family members in the 1970s, the potential solutions were rooted in mutual support and political activism with a focus on improving and strengthening services for individuals with severe mental illness living in the community. For advocates of the recovery model and policymakers in the 1990s and 2000s, reorienting services to further empower clients and encourage independence held the key to potentially helping individuals living with serious psychiatric disorders thrive in the community. Yet, as data and the experiences of Linda and Mila show, neither new mental health services nor new service philosophies have been able to adequately address the systemic clinical, societal and personal challenges brought about by deinstitutionalisation in California.

Acknowledgements The research presented in this chapter was conducted with generous support from the United States National Institute of Mental Health Grant R01MH080671; a Los Angeles County Department of Mental Health and University of California, Los Angeles Clinical and Translational Science Institute Translational Research Fellowship (CTSI Grant UL1TR000124); a Los Angeles County Department of Mental Health Stigma Reduction Grant; and a California Behavioral Health Center of Excellence Rapid Research Advances for Progress in Disparities Award.

CHAPTER 13

Callan Park in Transition

Roslyn Burge

> Callan Park is a community space with a history that bridges the old divides of institutional mental health and community physical health ... the whole of Callan Park is rediscovered as a therapeutic landscape.[1]

Introduction

Deinstitutionalisation in Australia has evolved slowly and the Stoller report of 1955 found that Australia lagged behind other nations in this respect.[2] International debates influenced Australia's response to changing community attitudes to mental illness and psychiatric care and a number of institutions in Australia and New Zealand have been examined by historians[3] and

[1] Leichhardt Council and McGregor Coxall (November 2011) *Callan Park Master Plan*, p. 166.

[2] A. Stoller and K.W. Arscott (1955) *Report on Mental Health Facilities and Needs of Australia* (Canberra, Commonwealth Department of Health).

[3] (2009) *Health and History, Special issue: Australian Asylums and Their Histories*, 11, 1 and (2003) *Health and History, Special issue: Histories of Psychiatry after Deinstitutionalisation: Australia and New Zealand*, 5, 2.

R. Burge (✉)
Sydney, Australia
e-mail: rburge@ozemail.com.au

D. Kritsotaki et al. (eds.), *Deinstitutionalisation and After*,
Mental Health in Historical Perspective,
DOI 10.1007/978-3-319-45360-6_13

documented in the plethora of reports issued by governments in the decades since the Second World War.[4] Sometimes government decisions about institutions appear to have been more keenly influenced by the value of public assets and effective management of agency needs than the supported delivery of care.[5] Overall, the challenges posed by deinstitutionalisation have been described as a bureaucratic nightmare.[6] Furthermore, a multiplicity of issues beyond the imperatives of best psychiatric care for individuals, or health department commitments has also influenced deinstitutionalisation.

This chapter examines the way deinstitutionalisation occurred at one hospital in New South Wales (NSW), Rozelle Hospital. It analyses how the financial value of the physical site influenced government policy and explores how the process of deinstitutionalisation was delayed and influenced by the community amid complex political, environmental, social and economic changes at local, state and federal level. The community's persistent use of the original name of the asylum, Callan Park, has sidelined its formal name and throughout this chapter 'Callan Park' is the collective name used for the whole site and 'Kirkbride' is used to describe the original nineteenth-century asylum complex at Callan Park.

The Sydney suburb, Rozelle, in which Callan Park is situated, was the formal name given to the new institution formed in 1976 when Callan Park Psychiatric Hospital and Broughton Hall Psychiatric Clinic, which took voluntary patients from 1921, were amalgamated. Together they form a largely intact Victorian landscape and are listed on the State Heritage Register (Fig. 13.1).[7] As places of high visibility and exceptional

[4] Successive reports include: Stoller and Arscott, *Report on Mental Health Facilities and Needs of Australia*; (1961) *Royal Commission of Inquiry in Respect of Certain Matters Relating to Callan Park Mental Hospital*; (1977) *Review of Mental Health Care*; D. Richmond (1983) *Inquiry into Health Services for the Psychiatrically Ill and Developmentally Disabled*, NSW Health; W. Barclay (1988) *Report to the Minister for Health*; B. Burdekin (1993) *Human Rights and Mental Illness: Report of the National Inquiry into the Human Rights of People with Mental Illness*; NSW Law Reform Commission (2013) 'People with Cognitive and Mental Health Impairments in the Criminal Justice System'. Between 2006 and 2012 reports of 30 separate statutory enquiries into mental health have been issued: S. Rosenberg (8 June 2015) 'Mental Health Crisis How Serious is it?' (Glebe, Politics in the Pub) https://www.youtube.com/watch?v=Jwo3uY5RdU4, accessed 4 March 2016.

[5] Community and Heritage Retention Services Unit, Property Services Group and CityWest Development Corporation were all NSW government entities. NSW Parliament, Legislative Council (20 October 1994) Joint Estimates Committees; and Asset Acquisition Program, Capital Charging policy, Shared Corporate Service Management Program and asset levies in D. Gates and R. Pye (n.d.c. 2006) *Emerging Asset Management Issues* (NSW Health Futures Planning Project).

[6] NSW Parliament (28 May 1997) Legislative Council.

[7] Callan Park was listed by the National Trust in 1974; on the Register of the National Estate in 1978; on the National Trust NSW's Endangered Places in 1999; on the Australian Garden History Society's Landscapes at Risk in 2016.

Fig. 13.1 Callan Park showing the Kirkbride Complex and proximity to Sydney, the Harbour Bridge and tip of the Opera House, February 2016 (Photographer: Jason Burcher; Retoucher: Thomas Ulman)

heritage value, their amalgamation was a significant beginning to the consolidation of psychiatric services. Callan Park was the most prestigious and costly government construction project across the nation when built in the 1880s but by the end of that century was already overcrowded, the very fate it had been designed to alleviate in other hospitals. It was among the largest institutions in the nation and had the largest number of admissions of the ten psychiatric hospitals in NSW (more than 3,000 each year between 1976 and 1982).[8] In 2007, a year before it closed, Callan Park still provided the largest number of public hospital psychiatric beds in the state.[9]

The landscape of Callan Park was pivotal to the government in the 1870s when selecting the site for a new asylum and is significant in understanding economic decisions about its closure and community opposition to its proposed disposal little more than a century later. Its open spaces, verdant hills and swales provide a natural environment and seclusion, a place designed to be therapeutic and a 'rare place of respite'.[10] It is a site of botanical riches with some of the tallest, oldest trees in the municipality, as well as a pocket of rainforest, and more tree species than Central Park in New York.[11]

Located on the harbour foreshore only six kilometres and six traffic lights from the city centre of the nation's largest city, Sydney, Callan Park covers 61 hectares (125 acres), the size of a small suburb. It is the largest open space in Leichhardt, the local government authority covering 11 square kilometres which embraces a number of suburbs, all with some part of their boundaries edged by Sydney Harbour.

Deinstitutionalisation at Callan Park

Deinstitutionalisation at Callan Park could be said to have begun informally with adverse publicity about the conditions and accommodation in which patients were housed, published in the press in 1947 and

[8] Richmond, *Inquiry into Health Services*, Table 1.

[9] Annual Report (2006/07) NSW Department of Health, unpaginated, 'Public hospitals activity levels'.

[10] Sydney Harbour Foreshore Authority (November 2007) *Callan Park Land Use Plan*, p. 21.

[11] Dean Simonsen (1 March 2011) *Tree Asset Register Callan Park Lilyfield* (Ringwood, Vic: Tree Logic Arboricultural Consultants), pp. 5–10; Zoe Baker (2012) *Callan Park History, Memory and Perceived (Healing) Space*, History Week, p. 1.

1948.[12] Agitated parliamentary debates in NSW ensued with references to potential select committees and royal commissions but it was not until the 1961 Royal Commission that conditions at Callan Park were formally examined.[13] It recommended improved conditions, functional subdivision of service delivery and a reduction in patient numbers. Despite the construction of new wards in the 1950s[14] many buildings were no longer adequate for patient care in the 1970s when state and federal governments began formal planning to deinstitutionalise, improve community care and sell off the large tracts of real estate on which psychiatric hospitals were usually located.[15]

A succession of reports and inquiries (including the Royal Commission, the Richmond Report of 1983, the state-wide review of healthcare services delivered to the psychiatrically ill and the intellectually handicapped in institutions and community-based facilities) recommended priorities for the development of new services, particularly community treatment, and the closure of mental hospitals.[16] After amalgamation in 1976 the first formal, large-scale planned rationalisation of Callan Park began in 1989. Between 1989 and 2007 the government produced five plans seeking to consolidate services, reduce patient and staff numbers, close the hospital and sell off portions of the land. A sixth plan was commissioned in 2010 by Leichhardt Council in response to state government inactivity which allowed the site to languish.[17] In 2015 the Minister for Heritage heralded a potential seventh sustainable plan.[18]

[12] In 1947 staff in eight psychiatric hospitals went on strike about the lack of suitable, separate accommodation for tuberculosis patients, *Newcastle Morning Herald and Miners' Advocate* (9 January 1947) 'Mental Hospital Staffs to Hold Stoppage', p. 3; *Tribune* (28 July 1948) 'Callan Park: Sun Stories "Grossly Exaggerated"', p. 6.

[13] (1961) *Royal Commission of Inquiry in respect of certain matters relating to Callan Park Mental Hospital* (13 September 1949) Legislative Assembly.

[14] Stoller Report, 'The Mental Health Needs of Australia', cited in NSW Department of Public Works and Services, Rozelle Hospital Site (15 February 1996) *Asset Management Strategy*, Infrastructure Partnerships, p. 12.

[15] C. Stuart (1978) Commonwealth Department of Health, 'Review of Mental Healthcare: A Discussion Paper', Policy and Planning Division, Canberra, Monograph series, No 2, cited in the Richmond Report, Appendices, Literature Review, p. 35.

[16] Richmond, *Inquiry into Health Services*. See also note 4 in this chapter.

[17] Brief for the Sydney Harbour Foreshore Authority Master Plan states, 'With the transfer of mental health facilities to Concord, Callan Park will lose the activity and tenure of its primary user. The site and its assets could then become vulnerable to deterioration.' p. 3.

[18] (1989) Master Development Control Plan; (1996) Asset Management Strategy; (1998) Scalabrini Retirement Home; (2002) Master Plan for the Rozelle Hospital site; (2007)

Few other psychiatric institutions in the nation have been so extensively (if repetitively and selectively) examined. Much of the publicly available documentation about the rationalisation of services at Callan Park has been prepared by NSW Planning (in conjunction with the Department of Health (DoH) and describes the site, services and constraints principally from a planning perspective. Each inquiry into patient care and conditions at Callan Park provided by psychiatric institutions in NSW in every decade in the second half of the twentieth century confirmed little progress in improving service delivery. Overcrowding remained a regular complaint, increased budgets were inadequate, patient treatment was not always restorative and patient numbers swelled and ebbed. These issues continue to preoccupy the contemporary debate about the 'missing middle', the too few alternatives between general practitioners and emergency departments for people needing psychiatric help.[19]

Although closing Callan Park was not envisaged in the 1940s, reducing patient numbers, improving care, returning patients to the community and closure of this site took six decades. Amid changing patterns of research and care, improved pharmaceutical management, the natural attrition of long-term patients and a gradual reduction in patient numbers in the latter decades of the twentieth century, the last patients were transferred to the newly constructed Concord Centre for Mental Health on 30 April 2008, when Callan Park closed.

In NSW there are five stand-alone institutions (17 nationally): some provide acute psychiatric services; some are repurposed within a general hospital; others have been closed to patients but have a new community of educational users or health bureaucrats occupying existing buildings.[20] Callan Park's buildings accommodated a mixed community of patients, bureaucrats and an educational campus from the 1980s.[21] A number of other communities are also associated with Callan Park: consumers, carers, staff, local residents, and recreational and informal users of the site.

Callan Park Land Use Plan; (2010) *Callan Park Your Plan*; correspondence from Minister for Heritage to Member for Balmain, 24 September 2015.

[19] S. Rosenberg (15 September 2015) 'From Asylums to GP Clinics: The Missing Middle in Mental Health Care', *The Conversation*, daily web-letter, https://theconversation.com, accessed on 4 March 2016; and 'Mental Health Crisis How Serious is it?'

[20] Rosenberg, 'Mental Health Crisis How Serious is it?; Bloomfield Hospital in Orange, Gladesville Hospital in Sydney.

[21] Wawina High School was established in the mid-1980s and Sydney College of the Arts in 1996.

Ascertaining patient responses to the closure of Callan Park or accounts of their experiences of treatment and accommodation at the site is difficult for members of the public.[22] Employment conditions restrict staff from speaking publicly about their work—the reverberations of both the 1961 Royal Commission and the 1990 sacking of a senior psychiatrist for speaking publicly to the media ensured collective silence among staff at Callan Park into the twenty-first century and beyond the hospital's closure.

Society's attitudes to mental health have changed but not the view of the local community which has overwhelmingly supported mental health services at Callan Park since 1989. Every survey (web or paper-based) of residents undertaken by Leichhardt Council confirms this view. Asked about Callan Park's future and uses the community asserted its wish for the site to remain accessible to the public and the continuation of psychiatric services. The council's 2001 survey found 98.6 per cent of the community wanted to retain mental health services. Another survey of 400 local residents in 2008 found 95 per cent were concerned, and 80 per cent were highly concerned, about the reduction in mental health services and facilities and wanted the council to preserve heritage and open space as well as improve existing mental health facilities.[23] This support underpinned the 2010 master plan. Since then Friends of Callan Park and the community have continued to discuss with government and consumer networks how best to activate the recommendations of that master plan's Mental Health Framework.

As deinstitutionalisation gained momentum and other social changes affected urban real estate in Sydney, the value of Callan Park increasingly represented a strategic asset for the government. The DoH initiated a concerted effort to capitalise on its large real estate portfolio throughout NSW and upgrade psychiatric services at a time when state finances were diminished. After an audit of state finances in 1988 disclosed unexpectedly high debt levels the government sought to intensify development in existing residential areas and on former industrial (often harbour foreshore) land

[22] In 2007 no consumers (and only two staff members) responded to emails circulated through the hospital (by and with the support of the General Manager) seeking participants in an oral history project. In 2015 only one consumer responded to an invitation to discuss consumer experiences of Callan Park circulated to more than 40 individuals by a well-respected consumer network.

[23] Leichhardt Council (20 August 2001) *Developing a Community Vision for Rozelle Hospital (Callan Park)*; Leichhardt Council (January 2008) *Community Survey Callan Park*, Piazza Consulting, pp. 6 and 13.

rather than greenfield sites on the metropolitan fringes.[24] The economics of asset valuation partly delayed deinstitutionalisation at Callan Park which was largely a planning/staffing process in the 1980s as society's attitudes to psychiatric care and to heritage shifted.[25]

Housing prices rose dramatically in the 1980s and reports to government highlighted the need for intensified urban consolidation.[26] The DoH published 'Blue Print for Health—A New Direction in Mental Health Services' in December 1988, and the following year began planning the Callan Park Master Development Control Plan, formally issued in March 1990 (Fig. 13.2). This plan proposed a centralised psychiatric service centre, a centre of excellence with congregated service delivery, a reduction in bed numbers from 389 to 278, patient accommodation in a newly constructed hospital providing services not possible in general hospitals or the community, and for acute patients unable to be accommodated in other hospitals.[27] Kirkbride was to close and 50 patients would be moved to new wards to be constructed at Callan Park. A further 20 patients were to be relocated to beds in community care, and 75 patients to beds in other hospitals. Some buildings were to be demolished, others refurbished for support services and Kirkbride adapted for reuse by the Sydney College of the Arts.

An additional benefit for the government was the sale of ten per cent of the land for asset realisation and medium density housing.[28] One statement repeated by government in all the plans was that a new psychiatric facility, much deserved by patients, could be constructed only if a portion of Callan Park was sold to pay for it. The community recognised ageing hospital wards were outdated but remained sceptical about government assurances that funds would be used for new psychiatric services.

[24] G. Searle (March 2007) 'Sydney's Urban Consolidation Experience: Power, Politics and Community', *Urban Research Program*, Research Paper 12, Griffith University, Brisbane; and T. Bonyhady (1995) 'The Battle for Balmain', *Urban Futures*, No. 18, 25–34.

[25] NSW Health (March 1996) *The Health of the People of New South Wales—Report of the Chief Health Officer*, Public Health Division, NSW Health Department, p. 78 in Fiona Manning, *Mental Health in NSW: Current Issues in Policy and Legislation*, NSW Parliament Briefing Paper No. 21/96.

[26] G. Searle (March 2007) 'Sydney's Urban Consolidation Experience: Power, Politics and Community', *Urban Research Program*, Research Paper 12, Griffith University, Brisbane; and T. Bonyhady (1995) 'The Battle for Balmain', *Urban Futures*, No. 18, 25–34.

[27] *Master Development Control Plan for Rozelle Hospital Sydney* (1989) Healthworks, NSW Public Works, p. 18, Rozelle Hospital Plan H5, Diagram 23; and p. 14.

[28] *Master Development Control Plan*, pp. 15, 138 and 142.

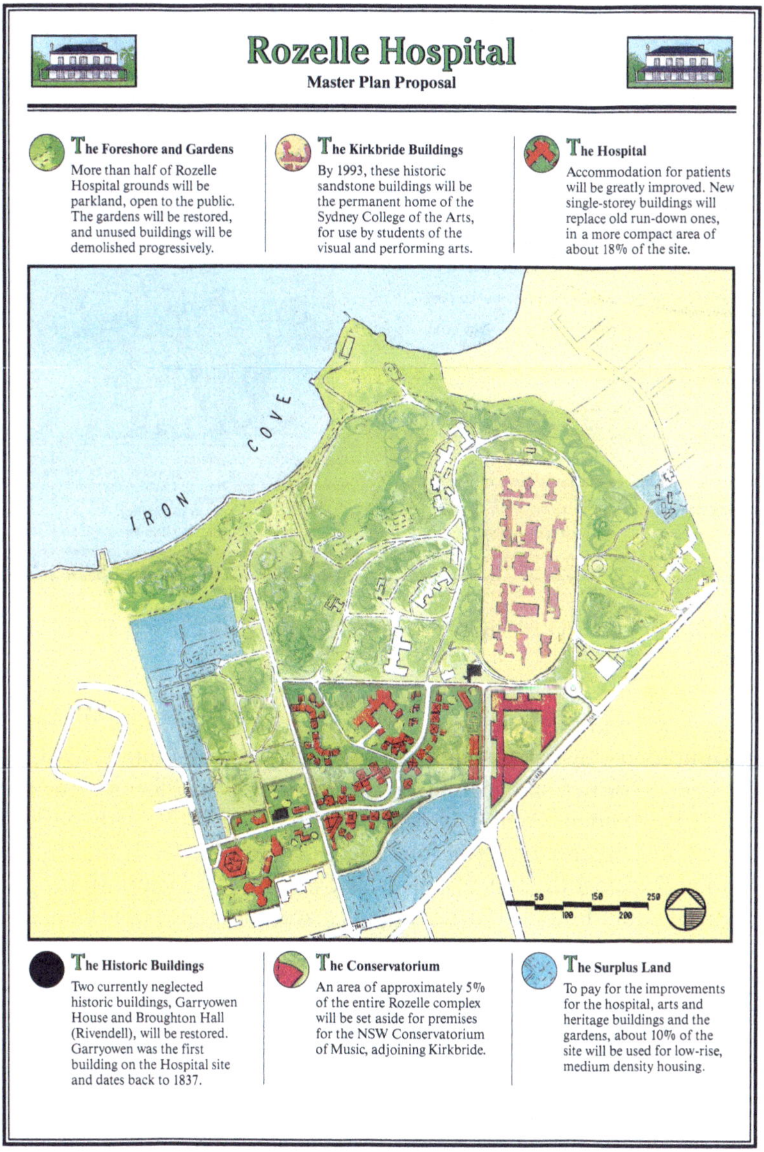

Fig. 13.2 Rozelle Hospital Master Plan, 1989 (Dr Alicia Lee Collection, Leichhardt Library)

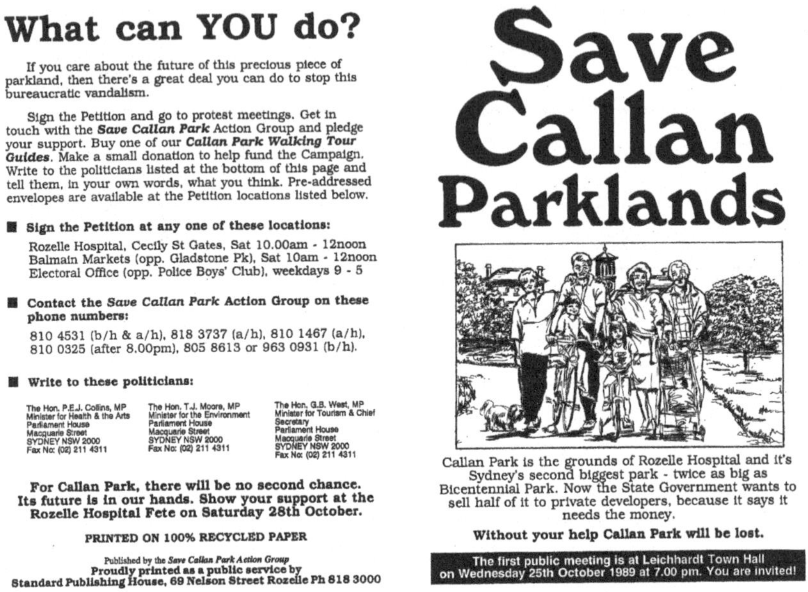

Fig. 13.3 Save Callan Parklands flyer, 1989 (Dr Alicia Lee Collection, Leichhardt Library)

Outcry greeted the 1989 plan. Staff were concerned about the impact of moving long-stay patients and job security. The community, led by Dr Alicia Lee, a former staff psychologist at Callan Park, formed the Save Callan Park Resident Action Group in September 1989, with staff psychiatrist and Director of Drug and Alcohol Services, Dr Jean Lennane, and lobbied against the loss of open space and beds.[29] Demonstrating the strength of community will to save open space, thousands of leaflets titled 'Save Callan Parklands' advertised a public meeting at Leichhardt Town Hall on 25 October 1989 (Fig. 13.3). Also, a petition entitled 'The Lungs of Leichhardt' requested Parliament to ensure that historic buildings and existing open spaces were retained in a municipality which ranked 37th out of 40 for open space per head.[30] Thirteen thousand people across

[29] Dr Lennane ignored employment conditions and repeated Health Department instructions, and continued to speak to the media about reductions in bed and patient numbers and was sacked in October 1990.

[30] Deputation to Minister for Health, Peter Collins, meeting notes (19 September 1989).

Sydney signed the petition, rebutting comments that it was 'nimbyism' and just a local issue.[31] In an open letter to residents the deputy leader of the opposition cited the dearth of information about the DoH's plans for Callan Park and the necessity for him (as Shadow Minister for Health) to apply for those plans under Freedom of Information legislation. He offered support for the campaign to save Rozelle Hospital from the government's 'mindless fire-sale mentality'.[32]

Dr Lee was no neophyte to campaigning and though she had worked at Callan Park she was still 'astounded at the degree of emotion which people express' about the site.[33] The Nurses' Association initiated industrial action opposing government plans in the 1980s but Save Callan Park Resident Action Group was the first organised community group which fought to save both mental health services and open space. Its membership included former staff members, local residents, and men and women from all walks of life. The group spearheaded community opposition for more than a decade and still functioned into the early 2000s.

Adding to the complexity of upgrading services and repurposing Callan Park as an arts precinct, the Minister for Health (also the Minister for the Arts) announced the Conservatorium of Music and its high school would be transferred from the city to Callan Park, requiring extensive revision of the 1989 plan. The Conservatorium's plans were more expensive than rebuilding in the city and the impact of planes flying low on approach to the airport along two parallel flightpaths directly above Callan Park was an obvious disturbance. Media coverage of these proposals highlighted community opposition, with the *Sydney Morning Herald* reporting 'Hostility simmering over threat to parkland' and the sale of open space 'to pay for a $17 million' hospital upgrade.[34]

[31] Nimbyism = not in my backyard. The community was labelled 'chardonnay-set' and accused of nimbyism when contesting the 2002 and 2007 master plans. In 2001, 17,000 signatures were collected in support of the Callan Park Trust Bill. Friends of Callan Park analysed a sample of 5,700; just over half were from the Leichhardt municipality, slightly less than half were from all over NSW and other states, and ten were overseas visitors.

[32] Dr Andrew Refshauge (25 October 1989) *Open Letter to the People of Balmain Lilyfield and Leichhardt.*

[33] Minutes (10 October 1989) meeting at Department of Health, re Rozelle Hospital, p. 3. Dr Lee had childhood polio (and was a member of the management committee of Polio NSW in the 1990s); fought local environmental and political battles in the 1960s; and in 1972 was a founding member of the NSW Women's Electoral Lobby, with other women who went on to take significant roles in academia and politics.

[34] *Sydney Morning Herald* (17 October 1989) 'Hostility simmering over threat to parkland'. That budget more than doubled within five years.

Working with the community, Leichhardt Council formed the Save all Rozelle Hospital Open Space Land Committee and at a public meeting passed a motion rejecting the use of Callan Park for anything other than hospital and open space purposes.[35] The DoH ignored community attitudes and proceeded with its plans. Some new wards were built but the extensive hospital reorganisation was abandoned and patients continued to be accommodated in buildings scattered across the site.[36]

Nevertheless the government continued to explore ways of rationalising large assets while maintaining a balance between maximum development and financial return, and community expectations about open space.[37] The Rozelle Hospital Site Asset Management Strategy of 1996 proposed five alternative plans for redeveloping psychiatric services, each providing 124 mental health beds.[38] Labelled 'extreme', one idea was to revisit an earlier proposal to close the hospital and relocate all patients to other sites, but there was no reference to mental health services, other than to note it was 'an important social issue'.[39] Confidentiality was a preoccupation of government and of 12 risks identified with closing the hospital or selling the land, community reaction and electoral backlash were viewed as the greatest threats—the plan was never released.

At the same time the DoH also began discussions in 1995 (or earlier) with the Scalabrini Brothers to sell the Catholic order two hectares of land at Callan Park for construction of a 200-bed retirement village. These discussions remained a tightly guarded secret for more than four years until

[35] *The Future of the Public Open Space at Rozelle Hospital* (leaflet 20 October 1989) included in Rozelle Hospital Site, *Asset Management Strategy* (December 1995) Planning Workshop *Background Notes*; Public meeting (15 July 1990), *The Glebe* (4 July 1990) Leichhardt Council public meeting notice; correspondence (February 1991) from Minister for Health to Minister for Family Services.

[36] The DoH budgeted $39.8 million in 1989 for the project. A development application was submitted to Leichhardt Council on 22 October 1990 and construction commenced in 1991. The 2010 master plan identified those wards for mental health services. One half have been renovated for NSW ambulance bureaucrats, the other half have remained empty since May 2008 because NSW DoH will not permit even nurses to be accommodated in these modern buildings.

[37] *Rozelle Hospital Site, Asset Management Strategy* (15 February 1996) NSW Department of Public Works and Services, Infrastructure Partnerships, p. 32.

[38] *Rozelle Hospital Site, Asset Management Strategy* (7 December 1995) Planning Workshop Outcomes, Appendix C, Rozelle Hospital Redevelopment Options. NSW Department of Public Works and Services; and Infrastructure Partnerships, 15 February 1996, p. 25.

[39] *Rozelle Hospital Site, Asset Management Strategy*, p. 9. Documentation about mental health planning was not publicised.

the community learnt a week beforehand that NSW Premier Bob Carr was to visit Callan Park on 28 November 1998 for the official sod-turning ceremony.[40]

The location was within the visual setting and curtilage of the exceptional heritage buildings of Kirkbride which the National Trust had recommended a quarter of a century earlier should not be built upon.[41] Once again the community was not consulted and was infuriated by the secrecy, the disregard of heritage and the about-turn by politicians. Indeed, Bob Car, when leader of the opposition, had vowed to use Parliament to block such sales when he launched his catalogue, 'A State for Sale', at the Save Rozelle Hospital stall at the Rozelle Hospital fete in 1989.

Almost immediately a public meeting of nearly 200 people rejected the proposal, demanded a new master plan and greeted the call to retain mental health services with a huge roar.[42] So began a 4-year campaign of press coverage and rallies, including 2,000 people marching in local streets, which hastened the possibility of election doom for the Labor Party.[43] Once more the DoH overlooked community views about public land which was not the exclusive possession of government, but 'part of the public estate, the endowment of the people'.[44]

The political landscape for Callan Park moved very fast in the new century. The DoH considered the viability of continuing mental health services at the site but the master plan released in May 2002 proposed the sale of nearly 20 hectares of Callan Park to accommodate 40 private residential buildings (of 1,020 home units), demolition of 50 buildings, and the creation of a public park.[45] The DoH, master planners and the local Member of Parliament all reiterated the need to sell part of the site

[40] Scalabrini Fathers—Site for Aged Persons Accommodation (4 June 1996): Memo from Tony Burg, Area Director—Northern, NSW DoH, Tab A2.

[41] James Semple Kerr (29 September 1974) *Report on Callan Park Rozelle, requested by the Health Commission of NSW*, National Trust of Australia (New South Wales) p. 2; the 'National Trust advocate (1) that no new buildings be erected on the escarpment to the north, or on the western slope of the gully to the east of the Kirkbride block'. Dr Kerr's recommendations were noted in the 1990 Plan, p. 16.

[42] Friends of Callan Park (n.d. January 1999) *Callan Park Saved*; Hall Greenland (September 2015) chair of the meeting, pers. comm.

[43] *Green Left Weekly* (11 September 2002) 'Save Callan Park!'

[44] H. Gilbert (January 2003) *Contested Public Lands: Values, Power and Public Process in Government Asset Disposal Programs—a Summary* (Brisbane: PRRES Conference), p. 2.

[45] Urban Design and Advisory Service (May 2002) *Draft Master Plan for the Rozelle Hospital site* for NSW DoH, pp. 5–6.

to fund construction of a new mental health centre to be built adjacent to Concord Hospital where all the services of a general hospital were available to consumers.[46]

Political risks foreshadowed in the government's strategy of 1996 became a reality on 19 October 2002 when public animus towards the NSW Labor government resulted in a federal by-election in the safe Labor seat of Cunningham in Wollongong (two hours south of Sydney) won by the Greens. Wollongong was the Labor heartland of steel and shipping industries; Balmain (adjacent to and within the same electorate as Callan Park) was the birthplace of the Australian Labor Party. Any similar loss at the 2003 state election would be untenable. The next morning the local Member of Parliament, Sandra Nori, announced the government had heard the people and would abandon the master plan: open space would be entrenched in public ownership forever and residential housing (of any kind) was ruled out.[47] Within three days the Treasury confirmed $43 million was suddenly available to fund construction of Concord Hospital.[48] Later that week when Ms Nori presented her private bill to Parliament, standing orders were suspended to ensure the swift passage of the Callan Park (Special Provisions) Act 2002, and two months later the legislation was signed on Christmas Eve. This site-specific legislation restricts uses to health, educational or community facilities, and structures can only be erected within the footprint and envelope of existing buildings.[49]

At the same time that tumult swirled around Callan Park the government established the NSW Parliament Select Committee of Inquiry into Mental Health Services in New South Wales in 2001. The Inquiry's report recommended, 'That NSW Health establish Rozelle Hospital as an asylum for the mentally ill, in the true meaning of the concept ... to provide medium to long-term rehabilitation services for people with a mental illness'.

[46] NSW DoH (January 2002) *Rozelle Hospital, Conservation Management Plan*, 1, p. 112. Concord Mental Health Centre is located approximately eight kilometres west of Callan Park.

[47] *Sydney Morning Herald* (21 October 2002) p. 6.

[48] *Future of Callan Park (Rozelle Hospital)* (29 March 2003) Correspondence from David Gates, Director Asset and Procurement Management, NSW DoH, to Dr B. Dunbar, Premier's Department.

[49] The Act was based on earlier legislation presented to Parliament on behalf of the community by the Greens in 1999.

The hospital grounds were to be identified as a health facility and not considered public space.[50]

'Haven' and 'asylum' have increasingly become part of debates reassessing the role of institutions. Stephen Garton and Oliver Sacks both asked in 2009 whether it is 'worthwhile reviving the classical ideal of asylum ... as one part of a larger comprehensive mental health care system', and whether the benign aspects of asylums have been forgotten.[51] Zoe Baker highlights the way memory and history of treatment have imprinted physically (in landscapes and gardens) and symbolically a sense of sanctuary to Callan Park.[52]

Psychiatric services for fewer than 150 patients were still being provided at Callan Park in 2007, though it was widely understood that the government was determined to close the hospital.[53] The Minister for Planning announced in June that year that in his role as administrator of the Callan Park legislation he had signed a memorandum of understanding with the University of Sydney and Sydney Harbour Foreshore Authority (a government authority which managed Callan Park between 2008 and 2015) to allow the university to explore an expanded campus presence at Callan Park, within the context of the legislation.[54]

The educational repurposing of Kirkbride began in 1996 but the scale of the university's $300 million construction envisaged in the 2007 *Callan Park Land Use Plan* meant dramatic developments not permitted under the Act. Having won this legislation, the community would not countenance any amendment. Meanwhile the university appeared not only to disregard the Act, but also to ignore the community's historic engagement to ensure that mental health services continued at Callan Park. At each workshop the university's buildings grew taller and statements about its 'prestigious image creating a strong positive identity' for the locality

[50] *Inquiry into mental health services in New South Wales* (2002) Legislative Council, Select Committee on Mental Health (Sydney), Final Report, Recommendation 10, p. 63.

[51] S. Garton (2009) 'Seeking Refuge: Why Asylum Facilities Might Still Be Relevant for Mental Health Care Services Today', *Health and History*, 11, 1, 28; O. Sacks (24 September 2009) 'The Lost Virtues of the Asylum', *The New York Review of Books*, 56, 14.

[52] Z. Baker, *Callan Park History, Memory and Perceived (Healing) Space*, History Week, p. 1.

[53] Three months after the Act was passed NSW DoH detailed Callan Park's liabilities and sought the Premier's Department's help 'decanting' mental health services from the site. Correspondence (29 March 2003) David Gates to Dr Bill Dunbar.

[54] Memorandum of Understanding between Minister Administering the Callan Park (Special Provisions) Act, The University of Sydney and Sydney Harbour Foreshore Authority, 28 June 2007, Item 4.6(b) and (c).

Fig. 13.4 Postcard opposing Sydney University plan, 2008 (Susan Dorothea White for Friends of Callan Park)

fanned community ire.[55] There was already a well-established identity and the public was dismayed that the minister formally responsible for the legislation would agree to any plan which potentially breached both the Act and the assurances given in Parliament by Sandra Nori in 2002: 'this legislation means Callan Park is protected forever and a day under a Labor government'.[56] Initially the council supported the university expansion but also called for retention of the hospital.[57] An intense campaign of rallies, media appearances and thousands of postcards flooding the offices of politicians and the vice-chancellor and senate of the university turned the tide (Fig. 13.4).[58] Once more politicians heard the people and the campus plan was abandoned on 22 October 2008.[59]

Customarily it is the responsibility of a site owner to conduct a master plan. Frustrated by government inertia since the hospital closed in 2008,

[55] Hassell (November 2007) *Callan Park—Land Use Plan*, Appendix C, *Social and Economic Assessment*, p. 9.

[56] NSW Parliament (24 October 2002) Legislative Assembly.

[57] *Village Voice* (July 2007) 'Uni lease shows "contempt" for community', pp. 1 and 4.

[58] 2,079 submissions were received, 67 per cent objecting to the proposal and 33 per cent supportive (though 'supporters were expressing support for the playing field'), General Purpose Standing Committee No. 4, (15 October 2008), cited in *Callan Park Your Plan*, Community Groups (undated).

[59] *Callan Park Secured for the future* (22 October 2008) News Release, Minister for Planning and Member for Balmain.

Leichhardt Council boldly commissioned the next master plan for Callan Park in 2010, reaffirming its position that Callan Park should be a Centre of Excellence in mental health. One of the key elements of the master plan was to review opportunities to reintroduce mental health activities to the site.[60] Professor Vaughan Carr was engaged to prepare the master plan's mental health discussion paper.[61] *Callan Park Your Plan* was the first community-led master plan for Callan Park, possibly the first master plan (anywhere) seeking the return of psychiatric services at a former institution, and the first plan in three decades to formally discuss and recommend the delivery of a new model of psychiatric services at Callan Park.

Delivering mental health services is complex, yet surveys of public opinion demonstrate community support for the adaptive reuse of Callan Park's buildings to deliver a cohesive set of diverse mental health social network services, and its landscape fulfils the therapeutic benefits cited in the literature review commissioned by the DoH in 2005.[62] The master plan reflects those aspirations for a new model of care and instead of a facility providing centralised services in a fixed formal entity, it proposes a range of services and social enterprises supporting mental health, including consumer-run enterprises, supported respite accommodation and wrap-around services.[63] Consumer groups such as the NSW Consumer Advisory Group support the return of mental health services (including step-up/down transitional accommodation) to the site, as do Professor David Copolov[64] and representatives from all sides of politics.[65]

It would be uncommon if everyone agreed to the closure of Callan Park or the return of a new model of care to the site. Some consumers are unable to return to the site because of their response to their care

[60] *Leichhardt Council Brief* (6 April 2010) 'Health Overlay 08' (including mental health) was one of 12 frameworks across all aspects of the site.

[61] Professor Carr is a Psychiatrist in the School of Psychiatry, University of NSW, and Chief Executive Officer and Scientific Director of the Schizophrenia Research Institute.

[62] C. Logan (2011) 'Preserving Health: Modern Hospitals as Historic Places', *APT Bulletin*, 42, 2/3, 47; NSW Department of Health (2005) *The effect of the built and natural environment on Mental Health Units on mental health outcomes and the quality of life of the patients, the staff and the visitors.*

[63] *Callan Park Master Plan Report* (November 2011) p. 166.

[64] Professor D. Copolov, AO, Pro Vice-Chancellor, Professor of Psychiatry and Honorary Professor of Physiology, Monash University, Professorial Fellow in the Department of Psychiatry, University of Melbourne.

[65] NSW Community Advisory Group (24 September 2010) *Position Statement, Planning the Future of Callan Park*, Submission to the Master Plan.

and experiences at Callan Park, while other patients, former staff and the community appreciate the beauty of the landscape and the place it plays in patient therapy. Phil Escott was a patient at Callan Park in the 1980s and regretted the move to Concord, explaining that 'the grounds were part of your working space—a rural retreat close to the city … and were more recuperative than the antidepressants'.[66]

Some professionals within the psychiatric field continued to refer to Callan Park's dark history and reject efforts to return services to the site.[67] A year after the Callan Park (Special Provisions) Act 2002 was passed *Australasian Psychiatry* published an article which questioned the community's intention to retain psychiatric services at Callan Park and public statements that the people own the site, not the Government.[68] Another psychiatrist's submission to the 2010 master plan identified Callan Park as a 'socially and emotionally contaminated' site.[69] Critics of the proposals to retain (and subsequently reintroduce) psychiatric services at Callan Park responded energetically to the way words such as 'institution', 'hospital', 'bed' and 'accommodation' were used by the general public, unaware of the medicalised nuances of these seemingly straightforward terms, but certain that the need for mental health services was imperative.

That general public included people concerned about the site, open space and heritage issues, as well as carers, and some consumers and former patients who cared about psychiatric services. These were not venerationalists seeking to preserve grand buildings or services set in aspic.[70] The community understands that Callan Park has a mixed history with memories of horror and abuse, and recollections of respite, care and recovery. However, every survey of the community's views has repeatedly indicated overwhelming support for mental health services and preservation of open space and heritage values at the site.

[66] Phil Escott, pers. comm. (25 February 2016). The Consumer Consultants model began at Callan Park in 1993. Mr Escott became a Consumer Consultant in 1999. He recalled some staff were shocked to learn the hospital was to move and there were differences of opinion about the benefits of the new Centre for Mental Health at Concord Hospital.

[67] S. Rosenberg (8 June 2015) 'Mental Health Crisis: How Serious is it?'

[68] A. Rosen and L. Manns (December 2003) 'Who Owns Callan Park? A Cautionary Tale', *Australasian Psychiatry*, 11, 4, 449.

[69] Mental Health Coordinating Council (14 January 2011) *A Contemporary Vision for Mental Health on the Callan Park Site—An Alternative to the Master Plan*, p. 11.

[70] L. McCarthy and P. Ashton (1996) 'Greening Oral History: Investigating Social Value in the Environment', *Oral History Association of Australia Journal*, 18, 47–51.

Callan Park has already demonstrated its potential as a place of community care. Houses once occupied by doctors' families were used successfully in the 1970s as short-term step-down accommodation for patients transiting back into the community. At the same time a review of long-stay consumers from Callan Park who moved to satellite housing and boarding houses in the community nearby was undertaken. Most described themselves as happier than they had been in Callan Park, yet without a programme of support the study found long-term consumers functioned at a low level irrespective of their accommodation, and living in the community became a 'ghetto existence'.[71] The need to augment services and support for consumers living in the community remains crucial.

Callan Park Your Plan was presented to the Minister for Planning on 11 November 2011 and confirmed that, 'The community has endorsed a plan ... which delivers a vision for a Wellness Sanctuary, bridging the gap between acute care and home life for those with mental illness, and contributing to the mental, physical and social health of the entire community.'[72] No decision has yet been made by government while further confidential reviews continue to be undertaken.

Friends of Callan Park was formed in 1998 in response to the Scalabrini development and has long advocated for modern, voluntary mental health services to be established at Callan Park.[73] It is not a mental health activist group as Milligan et al. describe, but a group, some with no knowledge or experience of mental ill health, yet all of whom have learnt about mental health issues throughout the campaign and remain committed because of their concerns about the removal of mental health facilities from the site.[74] Seeking to expand on the master plan framework, Friends of Callan Park and the Social Enterprise Mental Health Alliance developed the Callan Park Integrated Mental Health Precinct Model in 2012. The model proposes the site be part of a preventative care facility offering

[71] M. Flaherty, B. Flaherty and G. Parker (1977) 'A Comparative Study of Long-stay Psychiatric Patients Discharged to Boarding Houses and Satellite Houses', *Australian and New Zealand Journal of Psychiatry*, 11, 195.

[72] *Callan Park Master Plan* (November 2011) pp. 200 and 236.

[73] Friends of Callan Park was formed on 2 December 1998 and committed to work for the permanent safeguarding of Callan Park, including the retention of mental health services and open space.

[74] C. Milligan, R. Kearns and R. Kyle (2011) 'Unpacking Stored and Storied Knowledge: Elicited Biographies of Activism in Mental Health', *Health and Place*, 17, 7–16; K. Cruden (n.d.c. 2004) *Callan Park—A Terminal Case? 'More on the Findings'*, p. 2.

recovery and support services for people transiting into or out of acute mental healthcare, and emphasises that services should be consumer-centric and easily accessible to consumers and their carers. Less than one per cent of the state mental health budget was spent on specialised community residential services in 2011/2012. Callan Park seems an obvious solution and returning services described in the master plan would be relatively inexpensive for government.[75] The viability of this plan was patently clear to the inaugural NSW Mental Health Commissioner, John Feneley, when he met Friends of Callan Park less than a year after his appointment. Surprised the integrated model had received no traction he asked, 'Who are your enemies?'[76]

Conclusion

The recent past and planned deinstitutionalisation at Callan Park has been a confusing, complicated process. Community resistance to diminished services, relocation of patients and sale of public assets are some of the reasons cited for the delay in deinstitutionalisation. However, the DoH's plans to rapidly rationalise assets and eliminate risks without proper community consultation, and caution about safeguarding the local electorate's votes, augmented that delay.

Although the DoH viewed Callan Park as an economic asset requiring the best outcome for government, the community has repeatedly rejected this policy and since 1989 consistently asserted its collective wish to ensure psychiatric services are delivered at Callan Park. In his Annual Heritage Lecture in 2000, David Malouf described how such attachment overrides the 'rights and interests of the legal possessors' about a place in which the 'nation as a whole has a proprietary interest'.[77]

Deinstitutionalisation formally relocated all the patients and closed the hospital in 2008. Despite that reality, the community has continued to advocate for an integrated mental health services hub at Callan Park where its iconic, historic cultural landscape and buildings present opportunities for adaptive reuse. Statistics about the inadequacy of mental health services

[75] NSW Mental Health Commission (2014) *Living Well: A Strategic Plan for Mental Health in NSW* (Sydney), p. 55.

[76] Friends of Callan Park and Social Enterprise Mental Health Alliance (7 March 2013), meeting with Inaugural Mental Health Commissioner, John Feneley.

[77] D. Malouf (2000) *The National Trust Heritage Lecture* (Sydney).

across the nation provide compelling evidence to reassess the potential of the existing infrastructure at Callan Park which could provide an ideal environment for mental health consumers to work towards recovery, without stigma and with the support of a tolerant and accepting community. This represents a significant and unique opportunity for the establishment of an innovative and inclusive model of mental healthcare at Callan Park.

Index[1]

[1] Note: Page numbers followed by "n" denote notes.

D. Kritsotaki et al. (eds.), *Deinstitutionalisation and After*,
Mental Health in Historical Perspective,
DOI 10.1007/978-3-319-45360-6

The manufacturer's authorised representative in the EU is Springer Nature Customer Service Centre GmbH, Europaplatz 3, 69115 Heidelberg, Germany. If you have any concerns regarding our products, please contact ProductSafety@springernature.com

Printed and bound by CPI Group (UK) Ltd, Croydon, CR0 4YY
25/09/2026
02222859-0007